Use R!

Series Editors:
Robert Gentleman   Kurt Hornik   Giovanni Parmigiani

For further volumes:
http://www.springer.com/series/6991

Use R!

Series Editors:
Robert Gentleman   Kurt Hornik   Giovanni Parmigiani

For further volumes:
http://www.springer.com/series/6991

Jan Beyersmann • Arthur Allignol
Martin Schumacher

Competing Risks
and Multistate Models with R

Springer

Jan Beyersmann
Institute of Medical Biometry
and Medical Informatics University
Medical Center Freiburg
Freiburg Center for Data Analysis
and Modelling University of Freiburg
D-79104 Freiburg, Germany

Arthur Allignol
Institute of Medical Biometry
and Medical Informatics University
Medical Center Freiburg
Freiburg Center for Data Analysis
and Modelling University of Freiburg
D-79104 Freiburg, Germany

Martin Schumacher
Institute of Medical Biometry
and Medical Informatics University
Medical Center Freiburg
D-79104 Freiburg, Germany

ISBN 978-1-4614-2034-7   e-ISBN 978-1-4614-2035-4
DOI 10.1007/978-1-4614-2035-4
Springer New York Dordrecht Heidelberg London

Library of Congress Control Number: 2011941794

Printed on acid-free paper

Springer is part of Springer Science+Business Media (www.springer.com)

Preface

This book is about applied statistical analysis of competing risks and multi-state data.

Competing risks generalize standard survival analysis of a single, often composite or combined endpoint to investigating multiple first event types. A standard example from clinical oncology is progression-free survival, which is the time until death or disease progression, whatever occurs first. A usual survival analysis studies the length of progression-free survival only. A competing risks analysis would disentangle the composite endpoint by investigating the *time* of progression-free survival *and the event type*, either progression or death without prior progression. Competing risks are the simplest *multistate model*, where events are envisaged as transitions between states. For competing risks, there is one common initial state and as many target states as there are competing event types. Only transitions between the initial state and the competing risks states are considered.

A *multistate model* that is more complex than competing risks is the illness-death model. In the example of progression-free survival, this multistate model would also investigate death after progression. In principle, a multistate model consists of any finite number of states, and any transition between any pair of states can be considered.

This book explains the analysis of such data with R. In Part I, we first present the practical data examples. They come from studies conducted by medical colleagues where at least one of us has been personally involved in planning, analysis, or both. Secondly, we give a concise introduction to the basic concepts of hazard-based statistical models which is a unique feature of all modelling approaches considered. Part II gives a step-by-step description of a competing risks analysis. The single ingredients of such an analysis serve as key tools in Part III on more complex multistate models. Thus, our approach is in between applied texts, which treat competing risks or multistate models as 'further topics', and more theoretical accounts, which include competing risks as a simple multistate example. Our choice is motivated, firstly, by the outstanding practical importance of competing risks. Secondly, starting with competing risks allows for a technically less involved account, while at the same time providing many techniques that are useful for general multistate models.

The statistical concepts are turned into concrete R code. One reason for using R is that it provides for the richest practical toolbox to analyse both competing risks and multistate models. However, the practical implementation is explained in such a way that readers will be to able to, e.g., run Cox analyses of multistate data using other software, provided that the software allows for fitting a standard Cox model. Nonparametric estimation and model-based prediction of probabilities, however, are, to the best of our knowledge and at the time of writing, an exclusive asset of R.

The typical reader of the book is a person who wishes to analyse time-to-event data that are adequately described via competing risks or a multistate model. Such data are frequently encountered in fields such as epidemiology, clinical medicine, biology, demography, sociology, actuarial science, reliability, and econometrics. Most readers will have some experience with analysing survival data, although an account on investigating the time until a single, composite endpoint is included in the first two parts of the book. We do not assume that the reader is necessarily a trained statistician or a mathematician, and we have kept formal presentation to a minimum.

Likewise, we have refrained from giving mathematical proofs for the underlying theory. Instead, we encourage readers to use simulation in order to convince themselves within the R environment that the methodology at hand works. This *algorithmic perspective* is also used as an intuitive tool for understanding how competing risks and multistate data occur over the course of time.

Although refraining from a mathematically rigorous account, the presentation does have a *stochastic process flavor*. There are two reasons for this: firstly, it is the most natural way to describe multiple event types that happen over the course of time. Secondly, we hope that this is helpful for readers who wish to study more thoroughly the underlying theory as described in the books by Andersen et al. (1993) and Aalen et al. (2008).

How to read this book: The most obvious way is to start at the beginning. Chapter 1 presents the practical data examples used throughout the book. In Chapter 2, we recall why the analysis of standard survival data is based on *hazards*, and we then explain why the concept of a hazard is amenable to analysing more complex competing risks and multistate data. A further consequence is that the data may be subject to both the common *right-censoring*, where only a lower bound of an individual's event time may be observed, and *left-truncation*, where individuals enter the study after time origin. Such a delayed study entry happens, e.g., in studies where age is the time scale of interest, but individuals enter the study only after birth. The practical implications of Chapter 2 for competing risks are considered in Part II. Part III is on multistate models and frequently makes use of the competing risks toolbox.

Readers who urgently need to analyse competing risks data may proceed to the competing risks part of the book right away. They should at least skim over the description of competing risks as a multistate model in Chapter 3. The common nonparametric estimation techniques are in Chapter 4, and Cox-type regression modelling of the *cause-specific hazards* is explained in Section 5.2. These readers are, however, encouraged to read Chapter 2 later in order to understand why the techniques at hand work. In our experience, a practical competing risks analysis often raises questions such as whether the competing risks are independent or whether and when a competing risk can be treated as a censoring. Some of these issues are collected in Section 7.2. The theory outlined in Chapter 2 is necessary to clarify these issues.

Readers who wish to analyse multistate data in practice should have a clear understanding of competing risks from a multistate model point of view and as explained in detail in Part II. As stated above, this is so, because Part III frequently draws on competing risks methodology. The connection is that we are going to consider multistate models that are realized as a *nested sequence of competing risks experiments*; see Chapter 8.

This book is also suitable for *graduate courses* in biostatistics, statistics, and epidemiological methods. We have taught graduate courses in biostatistics using the present material.

The *R packages* and the *data* used in this book can be downloaded from the Comprehensive R Archive Network

> http://cran.r-project.org/

The book is also accompanied by web pages, which can be found at

> www.imbi.uni-freiburg.de/comprisksmultistate

The web pages provide the complete R code used to produce the analyses of this book as well as solutions to the Exercises. Sweave (Leisch, 2002) has been used to generate the LaTeX files of this book and to extract its R code. We also hope that readers will visit the web pages and leave us a message if they find any mistakes or inconsistencies.

We thank our medical colleagues who have granted us permission to use the data of their studies and to publish the data as part of R packages. This book has profited from collaborative work and/or comments from Adrian Barnett, Ronald Geskus, Nadine Grambauer, Stefanie Hieke, Aurélien Latouche, Reinhard Meister, Hein Putter and Christine Porzelius. We thank them all. Parts of this book have been written while the authors were supported by grant FOR 534 'Statistical modeling and data analysis in clinical epidemiology' from the Deutsche Forschungsgemeinschaft. This is gratefully acknowledged.

Freiburg,   *Jan Beyersmann*
                                                               Arthur Allignol
                                                            Martin Schumacher

Contents

Data examples and some mathematical
background

1

Data examples

In this book, we use both real and simulated data. One idea behind using simulated data is to illustrate that competing risks and multistate data can be conveniently approached from an algorithmic perspective. The data simulations are explained in their respective places in the book. In this section, we briefly introduce the real data examples. All of them are publicly available as part of the R packages used in this book.

Pneumonia on admission to intensive care unit, data set `sir.adm`

The data set is part of the `mvna` package. It contains a random subsample of 747 patients from the SIR 3 (*S*pread of nosocomial *I*nfections and *R*esistant pathogens) cohort study at the Charité university hospital in Berlin, Germany, with prospective assessment of data to examine the effect of hospital-acquired infections in intensive care (Wolkewitz et al., 2008). The data set contains information on pneumonia status on admission, time of intensive care unit stay and 'intensive care unit outcome', either hospital death or alive discharge. Pneumonia is a severe infection, suspected to both require additional care (i.e., prolonged intensive care unit stay) and to increase mortality.

The entry `sir.adm$pneu` is 1 for patients with pneumonia present on admission, and 0 for no pneumonia. A patient's status at the end of the observation period is contained in `sir.adm$status`, 1 for discharge (alive) and 2 for death. `sir.adm$status` is 0 for patients still in the unit when the data base was closed. These patients are called (right-) censored. A patient's length of stay is in `sir.adm$time`.

There were 97 patients with pneumonia on admission. Overall, 657 patients were discharged alive, 76 patients died, and 14 patients were still in the unit at the end of the study. 21 of the patients who died had pneumonia on admission.

The data set `sir.adm` is a competing risks example; that is, we investigate the time until end of stay *and* the discharge status, either alive discharge or hospital death. A challenge in the analysis of this data set is that pneumonia is found to increase the probability of dying in hospital, but appears to have no effect on the death *hazard*, i.e., loosely speaking, the daily probability of dying in hospital, given that one was still alive and in the unit at the beginning of the day.

We analyse the data set `sir.adm` in Sections 4.3, 5.2.2, 5.3.3, and Chapter 6.

Drug-exposed pregnancies, data set `abortion`

The `abortion` data set, shipped with the `etm` and `mvna` packages, contains information on 1186 pregnant women collected prospectively by the Teratology Information Service of Berlin, Germany. Among these pregnant women, 173 were exposed therapeutically to coumarin derivatives, a class of orally active anticoagulant, which are supposed to stop blood from clotting. Coumarin derivatives are vitamin K antagonists and are known to act as teratogens (i.e., they can disturb the development of an embryo or fetus). Controls consisted of 1013 women not exposed to potential teratogens. One aim of the study, which is discussed in detail by Meister and Schaefer (2008), was to assess the risk of spontaneous abortion after exposure to coumarin derivatives during pregnancy.

Women therapeutically exposed to coumarin derivatives have value 1 in `abortion$group`, which is 0 otherwise. Pregnancy outcomes are in `abortion$cause`, 1 for induced abortion, 2 for live birth, and 3 for spontaneous abortion. Study entry times are in `abortion$entry`, times of live birth or abortion are in `abortion$exit`. Pregnancy outcome is known for all women.

The data set `abortion` is a competing risks example; that is, we investigate the time until end of pregnancy *and* pregnancy outcome, either spontaneous abortion, induced abortion or live birth. A challenge in the analysis of this data set is that the time origin is conception, but women typically enter the study *after* conception. This is known as left-truncation. Women who, e.g., have a spontaneous abortion before their potential study entry time never enter the study.

Within the group of 173 exposed women, there were 43 spontaneous abortions, 38 induced abortions, and 92 live births. In the control group with 1013 women, there were 69 spontaneous abortions, 20 induced abortions, and 924 live births.

We analyse the data set `abortion` in Sections 4.4 and 5.2.2.

Cardiovascular events in patients receiving hemodialysis, data set `fourD`

The data set is part of the `etm` package and contains the control group data from the 4D study (Wanner et al., 2005). The background of the 4D study was that statins are known to be protective with respect to cardiovascular events for persons with type 2 diabetes mellitus without kidney disease, but that a potential benefit of statins in patients receiving hemodialysis had until then not been assessed. Patients undergoing hemodialysis are at high risk for cardiovascular events. The 4D study was a prospective randomized controlled trial evaluating the effect of lipid lowering with atorvastatin in 1255 diabetic patients receiving hemodialysis. Patients with type 2 diabetes mellitus, age 18–80 years, and on hemodialysis for less than 2 years were enrolled between March 1998 and October 2002. Patients were randomly assigned to double-blinded treatment with either atorvastatin (619 patients) or placebo (636 patients) and were followed until death, loss to follow-up, or end of the study in March 2004. The data set fourD contains those patients assigned to placebo treatment.

The 4D study was planned (Schulgen et al., 2005) and analysed (Wanner et al., 2005) for an event of interest in the presence of competing risks. The event of interest was defined as a composite of death from cardiac causes, stroke, and non-fatal myocardial infarction, whichever occurred first. The other competing event was death from other causes. Within the placebo group, there were 243 observed events of interest, 129 observed competing events, and 264 patients with censored event times. A patient's status at the end of the follow-up is in fourD$status. Possible values are 1 for the event of interest, 2 for death from other causes and 0 for censored observations. fourD$time contains the follow-up time.

We use the 4D data in the Exercises of Chapters 4, 5, and 6. The Exercises highlight how we can approach *real* data from an algorithmic perspective (Allignol et al., 2011c).

Bloodstream infections in stem-cell transplanted patients, data set okiss

The data set is part of the compeir package. It contains a random subsample of 1000 patients from ONKO-KISS (Dettenkofer et al., 2005); ONKO-KISS is part of the surveillance program of the German National Reference Centre for Surveillance of Hospital-Acquired Infections. KISS stands for Krankenhaus (Hospital) Infection Surveillance System. The patients in the data set have been treated by peripheral blood stem-cell transplantation, which has become a successful therapy for severe hematologic diseases. After transplantation, patients are neutropenic; that is, they have a low count of white blood cells, which are the cells that primarily avert infections. Occurrence of bloodstream infection (BSI) during neutropenia is a severe complication.

A patient's time of neutropenia until occurrence of bloodstream infection, end of neutropenia or death, whatever occurs first, is in time. A patient's status is in status, 1 for infection, 2 for end of neutropenia (alive and without prior bloodstream infection) and 7 for death during neutropenia without prior bloodstream infection. Patients censored while neutropenic have status equal to 11. Information on a patient's transplant type is in allo, which is 1 for allogeneic transplants and 0 for autologous transplants.

There were 564 patients with an allogeneic transplant. Of these, 120 acquired bloodstream infection. End of neutropenia, alive and without prior infection, was observed for 428 patients. These numbers are 83 and 345, respectively, for the remaining 436 patients with an autologous transplant. There were few cases of death without prior infection and few censoring events.

Autologous transplants are considered to be superior in terms of infection outcome. The challenge in this competing risks example is that autologous transplants in fact decreased the number of infections divided by the number of patients, but that they also increased the number of infections divided by the number of patient-days.

The ONKO-KISS data serve as a template for the simulated competing risks data that we analyse in Chapter 5. The data set okiss is used in the Exercises of Chapter 5.

Hospital-acquired pneumonia, data set icu.pneu

The data set is part of the kmi package and contains a random subsample of 1313 patients from the SIR 3 study described above; see the data set sir.adm. In contrast to sir.adm, the data set icu.pneu contains information on *hospital-acquired* pneumonia status, time of intensive care unit stay, and 'intensive care unit outcome', either hospital death or alive discharge. There is also additional covariate information on age and sex.

Hospital-acquired infections are a major healthcare concern leading to increased morbidity, mortality, and hospital stay. Length of stay is often used to quantify healthcare costs. Additional healthcare costs attributed to hospital-acquired infections are used in cost benefit studies of infection control measures such as isolation rooms.

Every patient is represented by either one or two rows in the data set. Patients who acquired pneumonia during intensive care unit stay have two rows. Each row represents a time interval from icu.pneu$start to icu.pneu$stop. On admission, all patients are free of hospital-acquired pneumonia. Their infection-free period is represented by the first data row. The infection status icu.pneu$pneu is 0. Patients with hospital-acquired pneumonia have a second data line which represents the time interval from pneumonia acquisition to end of stay or censoring. The infection status icu.pneu$pneu is 1 in the second data line. Observed end of stay at time icu.pneu$stop is indicated by icu.pneu$status equal to 1, which is 0 otherwise. For patients with two data lines, icu.pneu$status is always 0 in the first line. Finally, icu.pneu$event contains the hospital outcome, either 3 for death or 2 for alive discharge. The entry in icu.pneu$event has no meaning, if icu.pneu$status is zero.

21 observations were censored. 108 patients experienced hospital-acquired pneumonia. Of these, 82 patients were discharged alive and 21 patients died. Without prior hospital-acquired pneumonia, 1063 patients were discharged alive and 126 patients died.

The data set icu.pneu is a multistate example, namely a so-called illness-death model without recovery and with competing endpoints. All patients are in a common initial state on admission to the unit. As before, hospital outcome is modelled by competing endpoint states, but there is now also an intermediate state, which is entered by patients at the time that they acquire pneumonia. Interestingly, we again find that increased mortality after hospital-acquired pneumonia is mediated by delayed alive discharge. Compared to the data set `sir.adm`, however, an additional challenge arises through the time dependency of *hospital-acquired* infection status. E.g., in analyses of increased length of hospital stay, we must carefully distinguish between hospital days before and after the infection. The issue is that hospital stay before an infection must not be attributed to the infection.

We analyse the data set `icu.pneu` in Sections 9.2.1, 10.2.1, 11.1.3, and 11.2.3.

Ventilation in intensive care, data set `sir.cont`

The data set is part of the **mvna** package and contains another random sub-sample of 747 patients from the SIR 3 study described above; see the data set `sir.adm`. The data set `sir.cont` contains information on times of ventilation and time of intensive care unit stay. There is also additional covariate information on age and sex. Ventilated patients are supposed to require additional care, leading to prolonged intensive care unit stay.

`sir.cont` is an example of an illness-death model *with* recovery, because ventilation may be switched on and off during hospital stay. In addition, patients may either be on ventilation or off ventilation on admission to the unit. Events are represented in `sir.cont` in a multistate fashion. Switching ventilation *on* is represented by $0 \rightarrow 1$ 'transitions' as indicated in the columns `from` and `to`. Switching ventilation *off* is represented by $1 \rightarrow 0$ 'transitions'. Event times are in column `time`. Event times representing end of stay have value 2 in column `to`. The entry is `'cens'` for censored event times.

We investigate the impact of ventilation on length of stay in Sections 9.2.2, 10.2.2, and 11.4.

2

An informal introduction to hazard-based analyses

This chapter explains in a non-technical manner why methods for analysing standard survival data — one endpoint, observation of which is subject to right-censoring — transfer to more complex models, namely competing risks and multistate models, this book's topic.

In Section 2.1, we explain how right-censoring, where only an individual's minimum lifetime may be known due to closing of a study, say, leads to the hazard rate being the key statistical quantity. Probability estimates are then derived as deterministic functions of simpler estimators of the cumulative hazards. In fact, the technical difficulty of right-censoring is a consequence of an important conceptual aspect: time is not just another measurement on a scale, but plays a special role. Events happen over the course of time, e.g., illness often precedes death, and one has to wait in order to observe an event. This requires a dedicated statistical theory, and hazards are in general well suited to analyse events that occur over the course of time like survival data do. The concept of a hazard is also important for the data analyst's intuition: approximately, one may think of a hazard as the probability of experiencing an event within the next time unit conditional on presently being event-free. Say the event is death. Then this information may be more relevant given current vital status than an unconditional survival probability.

In Section 2.2, we explain why hazard-based techniques also apply to analysing competing risks data and multistate model data. Competing risks models allow for investigating different endpoint types that may occur at the event time in question. Occurrence of subsequent events may be investigated by multistate models. The type of multistate models that we consider are time-inhomogeneous Markov models, which are realized as a series of nested competing risks experiments. It is also discussed that these techniques allow for left-truncation in addition to right-censoring. Data are left-truncated if patients have a delayed study entry time. The mathematical basis behind these extensions are counting processes, which count different event types over the course of time, and martingales, which represent noise over the course of time.

The connection to counting processes and martingales is also explained in an informal way.

Finally, the brief Section 2.3 explains that asymptotic results are used for approximate inference when analysing event times in practice. Typically, the statistical techniques are of a nonparametric kind, and, e.g., using approximate normality has been found to work well even in moderate sample sizes.

Section 2.1 contains a lot of R code. Readers are encouraged to reproduce the code which explains how we may estimate cumulative hazards in the presence of right-censoring and how probabilities and their estimates may be computed from either the true or the estimated cumulative hazards. There is hardly any R code in Sections 2.2 and 2.3; this is what the remainder of the book is about.

2.1 Why survival analysis is hazard-based.

2.1.1 Survival multistate model, hazard, and survival probability

Figure 2.1 displays the simplest multistate model. An individual is in the initial

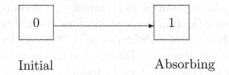

Initial   Absorbing

Fig. 2.1. Survival multistate model.

state 0 at time origin. At some later random time T, the individual moves to the absorbing state 1. 'Absorbing' means that the individual cannot move out of state 1, or that transitions out of state 1 are not modelled. Figure 2.1 is the classical model of survival analysis, if we interpret state 0 as 'alive' and state 1 as 'dead'. We are interested in the event time T; T is often called 'survival time' or 'failure time'.

To find out about T, we need to record data over the course of time, i.e. we need to record in which state, 0 or 1, an individual is for every point in time. This is what a stochastic process does. We write X_t for the state occupied by the individual at time $t \geq 0$, $X_t \in \{0, 1\}$. T is the smallest time at which the process is not in the initial state 0 anymore,

$$T := \inf\{t \,:\, X_t \neq 0\}. \tag{2.1}$$

This relationship between the stochastic process $(X_t)_{t\geq 0}$ and the event time T is illustrated in Figure 2.2. For illustration, consider an individual with event time $T = 52$. This individual will be in state 0 for all times $t \in [0, 52)$ and in

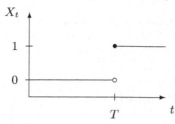

Fig. 2.2. Stochastic process $(X_t)_{t \geq 0}$ and the event time T: The bullet • is included in the graph, and the circle ○ is not.

state 1 for all times $t \geq 52$. Note that the state occupied at the event time T is the absorbing state 1, i.e., $X_T = X_{52} = 1$. This definition implies that the sample paths of the stochastic process, i.e.

$$[0, \infty) \ni t \mapsto X_t$$

are right-continuous, as illustrated in Figure 2.2. As state 1 is absorbing, data recording may stop for an individual with time T.

The statistical analysis of T is based on the hazard $\alpha(t)$ attached to the distribution of T:

$$\alpha(t) \cdot dt := P(T \in dt \mid T \geq t), \tag{2.2}$$

where we write dt both for the length of the infinitesimal (i.e., very small) time interval $[t, t + dt)$ and the interval itself. Equation (2.2) is a short, but more intuitive form of

$$\alpha(t) := \lim_{\Delta t \searrow 0} \frac{P(T \in [t, t + \Delta t) \mid T \geq t)}{\Delta t}.$$

Throughout this book, we assume that derivatives such as in Equation (2.2) exist. The hazard is 'just' a different 'representation' of the distribution of T:

Before we answer the question of why survival analysis is hazard-based, let us first note that knowing the cumulative hazard $A(t)$,

$$A(t) := \int_0^t \alpha(u) \, du, \tag{2.3}$$

suffices to recover the distribution function of T,

$$F(t) := 1 - S(t) := P(T \leq t) = 1 - \exp(-A(t)), \tag{2.4}$$

where $S(t) = P(T > t) = \exp(-A(t))$ is usually called the survival function of T. The right hand side of (2.4) is easily derived from (2.2) using standard calculus, but this adds little to our understanding. A more useful notion is product integration: because $dA(u) = \alpha(u)du$, we may rewrite (2.2),

$$1 - dA(u) = P(T \geq u + du \mid T \geq u). \tag{2.5}$$

The survival function should then be an infinite product over conditional probabilities of type (2.5). This is, in fact, the case. We call such an infinite product a product integral and write $\prod$. So,

$$S(t) = \prod_0^t (1 - dA(u)) \tag{2.6}$$

$$\approx \prod_{k=1}^K (1 - \Delta A(t_k)) \approx \prod_{k=1}^K P(T > t_k \mid T > t_{k-1}), \tag{2.7}$$

where $0 = t_0 < t_1 < t_2 < \ldots < t_{K-1} < t_K = t$ partitions the time interval $[0, t]$ and $\Delta A(t_k) = A(t_k) - A(t_{k-1})$. Now, the right hand side of (2.4) can simply be seen as a solution of the product integral in (2.6). The product integral itself, however, shows up again with the famous Kaplan-Meier or product limit estimator of the survival function, and, in a matrix-valued form, when we move from survival analysis to competing risks and multistate models. Let us now check the approximation of (2.6) by (2.7) empirically using R. The following function `prodint` takes a vector of time points and a cumulative hazard A as an argument and returns the approximation $\prod_{k=1}^K (1 - \Delta A(t_k))$.

```
> prodint <- function(time.points, A) {
+       times <- c(0, sort(unique(time.points)))
+       S <- prod(1 - diff(apply(X = matrix(times),
+                   MARGIN = 1, FUN = A)))
+       return(S)
+ }
```

A standard parametric example is the exponential distribution with constant hazard $\alpha(t) = \alpha$ and cumulative hazard $A(t) = \alpha \cdot t$. We exemplarily look at an exponential distribution with hazard 0.9,

```
> A.exp <- function(time.point) {
+       return(0.9 * time.point)
+ }
```

on the time interval $[0, 1]$:

```
> times <- seq(0, 1, 0.001)
> prodint(times, A.exp)
```

```
[1] 0.4064049
```

```
> exp(-0.9 * max(times))
```

```
[1] 0.4065697
```

The vector of time points does not have to be equally spaced:

```
> prodint(runif(n = 1000, min = 0, max = 1), A.exp)
```

[1] 0.4068678

A more flexible parametric model is the Weibull distribution with shape parameter θ and scale parameter γ. It has hazard $\alpha(t) = \gamma \cdot \theta \cdot t^{\theta-1}$ and cumulative hazard $A(t) = \gamma \cdot t^{\theta}$. Let us look at a Weibull distribution with scale 2 and shape 0.25,

```
> A.weibull <- function(time.point){
+    return(2 * time.point^0.25)
+ }
```

and the time interval $[0, 1]$ as before:

```
> prodint(times, A.weibull)
```

[1] 0.1234838

```
> exp(-2 * max(times)^0.25)
```

[1] 0.1353353

The approximation becomes better with an ever finer spaced partition:

```
> prodint(seq(0, 1, 0.000001), A.weibull)
```

[1] 0.1350193

The next section 2.1.2 explains why the cumulative hazard can still be estimated, if data are incomplete due to, e.g., individuals surviving the closing of a study. An estimator of the survival probability is then derived by computing the product integral with respect to the estimated cumulative hazard.

2.1.2 Estimation: The hazard remains 'undisturbed' by censoring.

The approximation (2.7) of the product integral (2.6), implemented via prodint, directly results in the Kaplan-Meier estimator of $S(t)$, if we substitute the increment of the cumulative hazard by an adequate estimator, which turns out to be the Nelson-Aalen estimator of the cumulative hazard. In other words, we may estimate the survival function S of the event time T by the product integral of an estimator of the cumulative hazard. This leads us to the question of why survival analysis is hazard-based, and, of course, how the cumulative hazard may be estimated: so far, we may estimate $S(t)$ either by the empirical survival function,

$$n^{-1} \cdot (\text{number of individuals surviving } t), \qquad (2.8)$$

if we start with n individuals at time origin. Or we may base things on estimating the cumulative hazard. A natural estimator of the increments $\Delta A(t)$ is

$$\Delta \widehat{A}(t) = \frac{\text{number of individuals failing at } t}{\text{number of individuals alive just prior to } t}, \tag{2.9}$$

so that we estimate the cumulative hazard as

$$\widehat{A}(t) = \sum_{k=1}^{K} \frac{\text{number of individuals failing at } t_k}{\text{number of individuals alive just prior to } t_k}, \tag{2.10}$$

if $0 < t_1 < t_2 < \ldots < t_{K-1} < t_K = t$ is the ordered sequence of observed failure times. It is a straightforward algebraic exercise to show that $\prod_0^t \left(1 - \mathrm{d}\widehat{A}(u)\right) = \prod_{k=1}^{K} \left(1 - \Delta \widehat{A}(t_k)\right)$ equals (2.8), if we observe the failure times for all n individuals. We briefly check this with R: we simulate 100 independent random variables from the exponential distribution with parameter 0.9,

```
> event.times <- rexp(100,0.9)
```

for which we wish to estimate the survival distribution at $t = 1$. We now need to compute the increments (2.9), which can be conveniently done using the **survival** package (Therneau and Grambsch, 2000): First, we create the fundamental 'survival object' using `Surv` on the simulated times. The `survfit`-function then gives us the necessary information:

```
> library(survival)
> fit.surv <- survfit(Surv(event.times) ~ 1)
```

Now, `fit.surv$time` is the vector of times $t_1 < t_2 < \ldots < t_{K-1} < t_K$, and `fit.surv$n.event` and `fit.surv$n.risk` are the numerator and denominator of (2.9), respectively. The function `A` computes (2.10):

```
> A <- function(time.point) {
+      sum(fit.surv$n.event[fit.surv$time <= time.point]/
+          fit.surv$n.risk[fit.surv$time <= time.point])
+ }
```

The estimator of the survival function at time 1 based on `A` and product integration then is

```
> prodint(event.times[event.times <= 1], A)
```

```
[1] 0.41
```

and the empirical survival function is

```
> sum(event.times > 1) / length(event.times)
```

```
[1] 0.41
```

The restrictive assumption required to use the empirical survival function (2.8) is that we are supposed to know the actual failure times of all individuals. This will usually not be the case. Event history data occur over the course of time, and a data analysis is regularly performed before or without knowing all failure times. E.g., a clinical study may be closed with, one hopes, many patients surviving, or individuals may drop out of a study because they move to a different place. In these instances, we will only know the minimum failure time. That is, we only know the actual failure time to be greater than a certain value, but not its precise value. This mechanism leads to incomplete observations and is known as (right-)censoring. In the presence of censoring, the empirical survival function (2.8) is rendered useless, as we cannot compute it anymore. *But hazards remain undisturbed by censoring:* recall Definition (2.2) of the hazard. Now introduce a censoring time C, independent of the event time T. (This is the so-called random censorship model.) The observation is

$$(T \wedge C, \mathbf{1}(T \leq C)), \tag{2.11}$$

where we write $\wedge$ for the minimum and $\mathbf{1}(\cdot)$ for the indicator function: $\mathbf{1}(T \leq C)$ equals 1, if T is less than or equal to C. $T \wedge C$ is the censored event time, and the event indicator $\mathbf{1}(T \leq C)$ tells us, whether $T \wedge C$ equals the actual event time T. Now, what is the probability of observing the actual event time in the small time interval $dt = [t, t+dt)$, conditional on the fact that neither event nor censoring have happened before t? I.e., what is $P(T \in dt, T \leq C \,|\, T \wedge C \geq t)$? The interval dt is so short that, assuming T and C to be different, at most one is in dt: if the event occurs in dt, it will be observed (still supposing $T \wedge C \geq t$). Because C and T are independent, the probability that the event occurs in dt, conditional on $T \wedge C \geq t$, is the same as in the absence of censoring:

$$\alpha(t) \cdot dt = P(T \in dt \,|\, T \geq t) = P(T \in dt, T \leq C \,|\, T \wedge C \geq t). \tag{2.12}$$

In words: censoring has not disturbed the hazard. As a consequence, we may estimate the cumulative hazard from censored data. Using product integration, this results in an estimator of the survival function. Equation (2.12) has farther reaching consequences, which we investigate in Section 2.2. These are also seen to be the reason why hazard-based techniques translate from the simple survival multistate model of Figure 2.1 to competing risks and more complex multistate models. Before we do so, let us briefly investigate estimation from censored data:

We say an individual with $T \wedge C \geq t$ is 'at risk' just prior to t. In order to estimate the cumulative hazard, adapting Equations (2.9) and (2.10) to the censored data set-up is straightforward:

$$\Delta \widehat{A}(t) = \frac{\text{number of individuals } observed \text{ to fail at } t}{\text{number of individuals } at \ risk \text{ just prior to } t}, \tag{2.13}$$

so that we estimate the cumulative hazard as

$$\widehat{A}(t) = \sum_{k=1}^{K} \frac{\text{number of individuals } observed \text{ to fail at } t_k}{\text{number of individuals } at \ risk \text{ just prior to } t_k}, \tag{2.14}$$

if $0 < t_1 < t_2 < \ldots < t_{K-1} < t_K = t$ is the ordered sequence of observed failure times. $\widehat{A}$ of (2.14) is the Nelson-Aalen estimator of the cumulative hazard. The product integral of $\widehat{A}$ is the Kaplan-Meier estimator of the survival function:

$$\widehat{S}(t) := \prod_0^t \left(1 - \mathrm{d}\widehat{A}(u)\right) = \prod_{k=1}^K \left(1 - \Delta\widehat{A}(t_k)\right) \tag{2.15}$$

We briefly revisit the R data example from above: in addition to `event.times`, we simulate censoring times, which we choose to be uniformly distributed on $[0, 5]$:

```
> cens.times <- runif(100,0,5)
```

The observable data are the censored event times $T \wedge C$,

```
> obs.times <- pmin(event.times, cens.times)
```

and the event indicator $1(T \leq C)$,

```
> event.times <= cens.times
```

The number of observed event times is

```
> sum(event.times <= cens.times)
```

```
[1] 80
```

We now have to refit the survival object, also telling `Surv` which event times were observed and which were censored:

```
> fit.surv <- survfit(Surv(obs.times,
+                        event.times <= cens.times) ~ 1)
```

The Kaplan-Meier estimator of the survival function at time 1 then is

```
> prodint(obs.times[obs.times<=1], A)
```

```
[1] 0.3967501
```

The result is reasonably close both to the estimate previously obtained in the absence of right-censoring and to the true value. Of course, we may also use the `survival` package in order to estimate the survival function at time 1:

```
> S <- fit.surv$surv
> S[fit.surv$time <= 1][length(S[fit.surv$time <= 1])]
```

```
[1] 0.3967501
```

So far, we have only estimated the survival function at one time point. Evaluating formula (2.15) at all observed event times yields an estimate of the survival *curve*. A plot of the estimated survival function together with its theoretical counterpart is displayed in Figure 4.3 in Section 4, where a more in-depth discussion of the Kaplan-Meier estimator, the `survival` package, and plotting the respective results is given.

In summary, survival analysis is hazard-based, because we can still estimate the cumulative hazard from right-censored data. We may then use product integration to recover the survival function or, equivalently, the distribution function. In the remainder of this chapter, we find that this program still works, in essence, with even more complex event data. Crucial to this is an intimate relationship between hazards and counting processes; the latter do a very intuitive thing: they count the number of observed events of a certain type over the course of time. However, the *interpretation* of hazard-based results becomes more involved with more complex event data, which is a major topic of this book.

2.2 Consequences of survival analysis being based on hazards

In Section 2.1, we illustrated that the analysis of event time data is based on hazards. This fact has a number of important consequences, which are briefly outlined below. In Section 2.2.1, we find that estimation of the cumulative hazard is intimately connected to counting processes and martingales. A counting process simply counts the number of observed events over the course of time. Martingale theory provides us with estimating equations and both small and large sample properties of estimators. This connection allows us to also analyse event time data which go beyond the right-censored, single-event type situation discussed in Section 2.1.

In Section 2.2.2, we show that the hazard-based approach can also account for left-truncated data, where patients have delayed study entry times. Sections 2.2.3 and 2.2.4 show how the current framework generalizes to competing risks and to time-inhomogeneous Markov multistate models. In addition to considering an event time, competing risks models also distinguish between different event types, one of which occurs at the event time. Multistate models can be thought of as being realized as a series of nested competing risks experiments: an individual may experience different events over the course of time, which are modelled as transitions between multiple states.

2.2.1 Counting processes and martingales

In Equation (2.12), we found that random right-censoring does not disturb the hazard. We reformulate (2.12) as

$$\mathrm{E}\left(\mathbf{1}(T \in \mathrm{d}t, T \leq C) \mid \mathrm{Past}\right) = \mathbf{1}(T \wedge C \geq t) \cdot \alpha(t)\, \mathrm{d}t, \qquad (2.16)$$

where 'Past' stands for knowledge about all failure or censoring events before t. Given the past, the at-risk indicator $\mathbf{1}(T \wedge C \geq t)$ is known. If the individual is at risk just prior to t (i.e., if we have that $T \wedge C \geq t$), the probability (conditional on the past) that we observe an event in the very small time interval $\mathrm{d}t$ is as in (2.12). However, if the individual is not at risk just prior to t because either a failure or a censoring event has happened before t (i.e., $\mathbf{1}(T \wedge C \geq t) = 0$), this probability is zero. This is summarized in (2.16).

As outlined earlier, Equation (2.12) implies that the cumulative hazard is estimable from the observable data. In fact, (2.12) shows that a key role in the estimation is played by the counting process

$$t \mapsto \mathbf{1}(T \leq t, T \leq C), t \geq 0, \qquad (2.17)$$

which has (infinitesimal) increments $\mathbf{1}(T \in \mathrm{d}t, T \leq C)$. The process (2.17) simply counts the number of observed events in the time interval $[0, t]$, either 0 or 1. Attached to the counting process is the at-risk process

$$t \mapsto \mathbf{1}(T \wedge C \geq t). \qquad (2.18)$$

The counting process may only jump (from 0 to 1) at time t, if the at-risk process equals 1. For estimation, we aggregate these processes over all individuals under study such that we count the number of observed events within the sample and over the course of time. The at-risk process then keeps track of the number of individuals currently at risk, i.e., without prior failure or censoring event. This is reflected in the Nelson-Aalen estimator (2.14).

Furthermore, Equation (2.12) is tantamount to the fact that

$$\mathbf{1}(T \leq t, T \leq C) - \int_0^t \mathbf{1}(T \wedge C \geq u) \cdot \alpha(u)\, \mathrm{d}u \qquad (2.19)$$

is a so-called martingale. Martingale theory provides a powerful tool to derive estimators and test statistics as well as to study their small and large sample properties. The latter may be used for approximate inference in practice. An in-depth treatment of the application of martingale theory to the analysis of event time data is beyond the technical level of this book. Interested readers are referred to Andersen et al. (1993) and Aalen et al. (2008). We, however, often make use of the results provided by the application of martingale theory. E.g., variance estimators and approximate 95% confidence intervals may conveniently be derived in this way.

We note, though, that equations (2.16) and (2.19), aggregated over all individuals under study, suggest estimating $A(t) = \int_0^t \alpha(u)\, \mathrm{d}u$ by the Nelson-Aalen estimator

$$\widehat{A}(t) = \sum_{k=1}^K \frac{\text{number of individuals observed to fail at } t_k}{\text{number of individuals at risk just prior to } t_k},$$

where the summation is over all event times t_k, which are less than or equal
to t (cf. (2.14)). Equation (2.16) also suggests that $\widehat{A}$ should be an almost
unbiased estimator of A, as long as the risk set (i.e., the set of all individuals
currently at risk) is non-empty with a high probability. Martingale theory can
be used to show that this is actually the case.

In other words, the martingale (2.19), potentially aggregated over all individuals, can be considered as a noise process. Figure 2.3 shows the counting
process

$$t \mapsto \text{number of individuals observed to fail in } [0, t]$$

computed based on the simulated data in Section 2.1 and its so-called compensator

$$t \mapsto \int_0^t (\text{number of individuals at risk just prior to } u) \cdot 0.9 \, \mathrm{d}u,$$

i.e., the integral over the at-risk process times the hazard $\alpha(t) = 0.9$ of the
uncensored event times (cf. equations (2.16) and (2.19)). Figure 2.3 illustrates

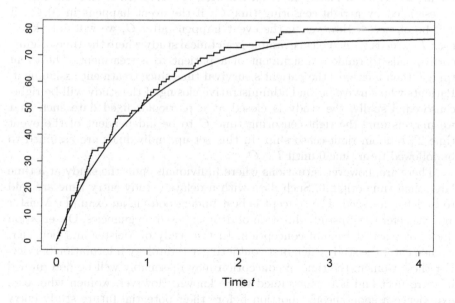

Fig. 2.3. *Simulated data.* The step function is the counting process of observed
events. The smooth line is the compensator of the counting process, i.e. the integral
over the risk set times the true hazard. Note that the compensator is only an almost
smooth line which has 'edges'. The left hand derivative does not equal the right
hand derivative at time points where the number of individuals at risk changes.

that the counting process approximates the compensator, which is intimately related to the cumulative hazard. The martingale itself, i.e., the counting process minus the compensator, is unpredictable zero-mean noise.

Readers are encouraged to check the approximation illustrated in Figure 2.3 in R: as explained in Section 2.1, the increments of the counting process at times `fit.surv$time` are contained in `fit.surv$n.event`, which allows for convenient computation of the counting process itself. The integral over the at-risk process times 0.9 may be computed from `fit.surv$n.risk`, which contains the number of individuals at risk just prior to times `fit.surv$time`.

The fact that counting the number of observed events over the course of time approximates a quantity which is closely related to the cumulative hazard, which, in turn, is a key target quantity for estimation, makes counting processes and martingales a starting point for a rich statistical theory. Sections 2.2.2–2.2.4 discuss how we can profit from the counting process approach for more complex event time patterns.

2.2.2 Left-truncation and right-censoring

So far, we have considered the situation where observation of an event time T is restricted by a right-censoring time C: if the event happens in $(0, C]$, it will be observed. However, if the event happens after C, we will only know that T exceeds C. A typical example is a clinical study where the time origin 0 corresponds to random assignment of a patient to a treatment. The event time T then measures the patient's survival time since treatment assignment. Patients who survive beyond administrative closing of the study will be right-censored. Usually, the study is closed at a particular fixed date such that we may assume the right-censoring time C to be independent of the event time T, random right-censorship. In this set-up, individuals are assumed to be followed from time 0 until $T \wedge C$.

There are, however, situations where individuals enter the study at a time later than time origin 0. Such data with a delayed study entry time are said to be left-truncated. The concept is best understood via an example. Meister and Schaefer (2008) study duration of drug-exposed pregnancies. Observation does not start at time of conception. In the study of Meister and Schaefer, women enter the study when first contacting a Teratology Information Service. For these women, the time of conception may reasonably well be determined in retrospect and is thus assumed to be known. However, women who, e.g., experience a spontaneous abortion before their potential future study entry time never enter the study. The hazard/counting process based approach of Section 2.2.1 also allows us to analyse such left-truncated data. The data analysed in Meister and Schaefer (2008) are available in the R package `etm`.

In addition to T and C, we denote an individual's left-truncation/study entry time by L: the event will only be observed, if it happens in $(L, C]$. If it happens in $(L, C]$, we will know T. An individual who experiences an event before its left-truncation time (i.e., $T \leq L$) will never enter the study. An individual under study (i.e., an individual with $L < T$) is right-censored, if it experiences an event after its right-censoring time (i.e., $C < T$). Data subject to right-censoring only are included in this set-up: they formally have a left-truncation time $L = 0$. Similarly, data which are only subject to left-truncation formally have a right-censoring time beyond the largest possible event time.

Potential subsequent occurrences of L, T, and C are schematically illustrated in Figure 2.4. The individual in Figure 2.4 a) enters the study at time L

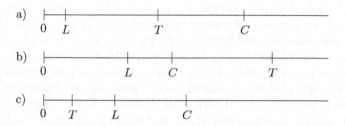

Fig. 2.4. a) Individual with an observed event. b) Censored individual. c) Individual with an event before study entry.

after time origin 0. The individual's event is observed at time T, because the censoring time C is larger than T. The individual in Figure 2.4 b) also enters the study at a time L, but the observation of the individual is right-censored. In Figure 2.4 c), the individual experiences an event before its study entry time, i.e., $T < L$. This individual never enters the study.

We assume for the moment random right-censoring and random left-truncation: T is independent of (L, C). Then Equation (2.16) generalizes to

$$E\left(1(T \in dt, L < T \leq C) \mid \text{Past}\right) = 1(L < t \leq T \wedge C) \cdot \alpha(t)\,dt, \qquad (2.20)$$

where 'Past' now means knowledge about all failure, truncation, or censoring events before t. As in (2.16), Equation (2.20) states that the probability of observing an event in dt is 0, if the individual is not at risk just before t. However, if the individual is at risk, an event that happens in dt will be observed; such an event occurs with probability $\alpha(t)\,dt$. The key point is that the at-risk process

$$t \mapsto 1(L < t \leq T \wedge C) \qquad (2.21)$$

now also accounts for left-truncation. An individual is only 'at risk' right after the time L of study entry. The Nelson-Aalen estimator (2.14)

$$\widehat{A}(t) = \sum_{k=1}^{K} \frac{\text{number of individuals observed to fail at } t_k}{\text{number of individuals at risk just prior to } t_k},$$

of the cumulative hazard $A(t) = \int_0^t \alpha(u)\,\mathrm{d}u$ is straightforwardly adapted to data subject to both left-truncation and right-censoring. The denominator now includes all individuals who have entered the study before t_k but have not experienced an actual event or a censoring event before t_k. The numerator counts the number of observed events at t_k within the set of these individuals.

A further important consequence of Equation (2.20) is that it allows us to relax the assumption of random right-censoring and random left-truncation. The application of martingale theory and counting processes only requires Equation (2.20) to hold, but not necessarily random right-censoring or random left-truncation. Restrictions of observing T that fulfill (2.20) are known as independent right-censoring and independent left-truncation, respectively. A crucial issue here is that L and C may depend on the 'Past': if covariates are considered in the statistical analysis, L and C may depend on past covariate values. A simple example of this is a clinical study where censoring may differ between treatment groups; see, e.g., Clark et al. (2002). A more complex example is the illness-death model as considered in Section 2.2.4 below. In this model, individuals may experience different events over time, i.e., undergo 'healthy' ↔ 'diseased' transitions and may also die. A right-censoring mechanism that is independent censoring in the aforementioned sense may depend on whether an individual is currently 'healthy' or 'diseased'. Because of this potential dependency, however, such right-censoring would not be random any more.

So far, our discussion of left-truncation and right-censoring referred to situations where observation of an individual was restricted due to some 'external' mechanism. Observation does not start before the time of left-truncation, and an event that happens after the right-censoring time will not be observed. In Sections 2.2.3 and 2.2.4, our aim is to estimate multiple (cumulative) hazards that correspond to multiple event types. Equation (2.20) can be generalized to such situations. A key point is to adapt the risk sets. In the aforementioned illness-death model, individuals will only be at risk of making a 'diseased' → 'dead' transition, after having acquired the disease and thus having entered the disease state. The set of individuals at risk of making a 'healthy' → 'diseased' transition not only excludes individuals who have previously fallen ill and have not yet recovered, but also those who previously died without prior disease. As discussed below, the appropriate changes to the risk set may be made by coding these as left-truncation and right-censoring, respectively, but these modifications of the risk set are due to the presence of multiple event types and not to external restrictions on an individual's observable data. We note, however, that, conversely, the presence of multiple event types may motivate data collection which is subject to left-truncation. E.g., in hospital epidemiology, the time scale of interest typically is time since admission to hospital, but sometimes data are collected conditional on detection of some infectious strain during hospital stay (Beyersmann et al., 2011).

In brief, left-truncation corresponds to observation being switched on, and right-censoring corresponds to observation being switched off. If these switches, which are, in fact, a 'censoring' process, are independent as explained above, 'the Nelson-Aalen estimator works'. We make two final comments on this: First, the idea of switching observation on and off may be generalized to quite complex observation schemes called filtering, and it disposes of the latent variables L and C. We do not further pursue the concept of filtering here. The variables L and C are latent in the sense that, e.g., L is unobservable if $L \geq T$, which is somewhat unpleasant. In contrast, the concept of 'observation on' as mirrored in the risk set does not require these latent times. In fact, our discussion of the illness-death model above has implicitly used the concept of 'observation on/off' rather than latent times. We have, however, chosen to use L and C in line with many accounts in the applied literature. Second, the independence assumption is obviously crucial, and it would therefore be useful if one could check it for a real data set. Unfortunately, there are identifiability problems that typically prevent checking the assumption for right-censoring, but it may be investigated for left-truncation. For the latter, see Section 11.3.

We finally note that left-truncation should not be confused with left-censoring. For a left-censored event time, we know that an event has happened before a left-censoring time in the past, but the exact time is not known. E.g., the occurrence time of a certain disease is left-censored, if it is only known to have occurred before the time of diagnosis. Klein and Moeschberger (2003) give a very readable account of the different variants of truncation and censoring. Our current approach conveniently allows for right-censoring, which is the most frequent reason for incompletely observed event time data, and left-truncation.

2.2.3 Competing risks

So far, we have considered a time T until one single possible event. The standard example is time until death, hence the name survival analysis. Often, however, a combined endpoint is considered. E.g., medical studies often investigate 'disease-free survival', i.e., time until (recurrence of a) disease or death (without prior disease), whatever comes first. In economics, one might wish to study durations of unemployment, ended either by finding a new job or retirement. Thus, T in general denotes time until some first event. The aim of a competing risks model is to distinguish between the possible types of that first event.

The analysis of competing risks is covered in depth in Chapters 3–7. At the current stage, the key question is how to generalize the basic two-state survival model of Figure 2.1 to competing risks. Our preceding discussion implies that an individual moves into the absorbing state of Figure 2.1 at time T, when the first of the possible events under study occurs. In other words, the absorbing state of Figure 2.1 represents a combined endpoint. The two-state survival model may now be generalized to competing risks by introducing several competing absorbing states which represent the possible event types. Occurrence of a competing event is modelled by a transition into the corresponding competing event state. Such a model is depicted in Figure 2.5 and a finite number J of competing risks. Figure 3.1 in Chapter 3 displays the corresponding model for two competing risks.

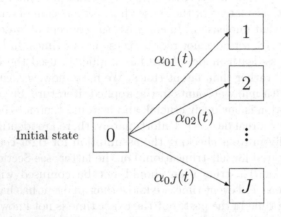

Fig. 2.5. Competing risks multistate model with cause-specific hazards $\alpha_{0j}(t)$, $j = 1, 2, \ldots, J$. The vertical dots indicate the competing event states $3, 4, \ldots, J - 1$.

Restricting for the moment the discussion to two competing risks, the stochastic process $(X_t)_{t \geq 0}$ attached to Figure 2.1 may easily be extended to the competing risks setting. Again, X_t simply denotes the state occupied by the individual at time $t \geq 0$. X_t equals 0, if the individual is still event-free at time t. Coding, as in Figure 3.1, the two potential competing events as 1 and 2, X_t equals 1, if event type 1 has occurred in $[0, t]$. If event type 2 has occurred in $[0, t]$, $X_t = 2$. As before, the event time T is the smallest time at which the process is not in the initial state 0 anymore; $T := \inf\{t : X_t \neq 0\}$ (cf. Equation (2.1)).

In addition to the event time T, competing risks data consist of a second component, the event type. Recall from our discussion of Figure 2.2 that X_T denotes the absorbing state entered at time T. In our setting with two competing event states, X_T equals either 1 or 2. I.e., X_T denotes the event type, and complete competing risks data consist of the tuple (T, X_T). More than two competing risks are easily included in this set-up by letting $X_T \in \{1, 2, \ldots, J\}$.

As illustrated in Figure 3.1, we now have one event-specific hazard per competing event. Paralleling Definition (2.2), these are defined as

$$\alpha_{0j}(t) \cdot dt := P(T \in dt, X_T = j \,|\, T \geq t), \, j = 1, \ldots, J, \qquad (2.22)$$

where the index $0j$ denotes the transition type out of the initial state 0 into the competing event state j. The α_{0j}s are often called cause-specific hazards (e.g., Prentice et al., 1978). The interpretation of (2.22) is that $\alpha_{0j}(t) \cdot dt$ is the probability that a type j event happens in the small time interval $dt = [t, t + dt)$, conditional on the fact that no event (of any type) has happened before t. For the more complex multistate models considered in Section 2.2.4 below, it is useful to rewrite Definition (2.22) in terms of the simple stochastic process $(X_t)_{t \geq 0}$,

$$\alpha_{0j}(t) \cdot dt = P(X_{(t+dt)-} = j \,|\, X_{t-} = 0), \, j = 1, \ldots, J, \qquad (2.23)$$

where X_{t-} denotes the state occupied just before time t.

As in Section 2.2.2, we call mechanisms of left-truncation and right-censoring independent, if they do not change these probabilities, i.e.,

$$E\left(1(T \in dt, X_T = j, L < T \leq C) \,|\, \text{Past}\right) = 1(L < t \leq T \wedge C) \cdot \alpha_{0j}(t)\, dt, \qquad (2.24)$$

$j = 1, \ldots, J$. A cause-specific Nelson-Aalen estimator of the cumulative hazard $A_{0j}(t) = \int_0^t \alpha_{0j}(u) \, du$ is now given as

$$\widehat{A}_{0j}(t) = \sum_{k=1}^{K} \frac{\text{number of observed type } j \text{ events at } t_k}{\text{number of individuals at risk just prior to } t_k}, \qquad (2.25)$$

$j = 1, \ldots, J$, where the summation is over all event times t_k, which are less than or equal to t. In Chapter 3, we show that we can compute probability estimates as deterministic functions of $t \mapsto (\widehat{A}_{01}(t), \ldots, \widehat{A}_{0J}(t))$. Here, we note two important facts which have already been alluded to earlier: first, the numerator in (2.25) now represents increments of a cause-specific counting process. Second, the risk set in (2.25) excludes all prior type j events, all prior censoring events, and all prior events of a type $\tilde{j}$, $\tilde{j} \neq j$.

In other words, when coding computation of $A_{0j}(t)$, say, we may code type $\tilde{j}$ events, $\tilde{j} \neq j$, as censoring events and only count type j events as 'actual events': occurrence of type $\tilde{j}$ events acts as independent right-censoring with respect to type j events. This means that removal of prior type $\tilde{j}$ events from the risk set allows for estimation of $A_{0j}(t) = \int_0^t \alpha_{0j}(u) \, du$. However, in Chapter 3, we also show that such 'censoring by a competing event' is informative in the sense that probability estimates depend on computing all $\widehat{A}_{01}(t), \ldots, \widehat{A}_{0J}(t)$. This has two important implications for any competing risks analysis.

- In a cause-specific hazards analysis, competing events may be coded as a censoring event.
- This has to be done for every competing event type in turn.

These two steps are, in particular, illustrated in Chapter 5.

Finally, in Section 2.2.4 below, we extend the hazard-based approach to multistate models, which can be thought of as being realized as successive nested competing risks experiments. To this end we note that competing risks data can be considered as realizations of a two-step simulation experiment that determines the time T at which the event occurs via the all-cause hazard $\alpha(t) = \alpha_{01}(t) + \dots \alpha_{0J}(t)$ (i.e., the usual hazard of the event time T); the event type X_T for a given time T is determined via a multinomial experiment that decides with probability $\alpha_{0j}(T)/\alpha(T)$ on $X_T = j$. This simulation point of view towards competing risks data shows up again and again in the main part of the book. See, in particular, Sections 3.2 and 5.2.2 for an in-depth treatment.

2.2.4 Time-inhomogeneous Markov multistate models

The preceding Section 2.2.3 on competing risks generalized the standard survival set-up with one event time T and one event type to modelling different possible event types (the competing risks) that may occur at time T. I.e., competing risks model time until some first event and the type of the first event, but (by definition) potential subsequent events are not modelled. Multistate models allow for modelling both the occurrence of different event types and the occurrence of subsequent events, the latter potentially of different types.

The present section is organized as follows. We first consider some important examples of multistate modelling. In such a model, events are modelled as transitions between different states. Next, we explain that a sequence of events/transitions (i.e., a realization of a multistate process) can be thought of as being realized as a series of nested competing risks experiments. This implies that the estimation techniques of Sections 2.2.1 and 2.2.2 also work in the more complex multistate situation. We then discuss that such a multistate model is time-inhomogeneous Markov and introduce its transition probabilities and transition hazards. Next, the Nelson-Aalen estimator of the cumulative transition hazards is considered, and finally we show how matrix-valued product integration yields the matrix of transition probabilities as a deterministic function of the cumulative transition hazards. Replacing the cumulative transition hazards by their Nelson-Aalen estimators results in the Aalen-Johansen estimator of the transition probabilities.

Examples of a multistate process

In Section 2.1, we explained that we need to keep track of an individual's status over the course of time, which is what a stochastic process $(X_t)_{t \geq 0}$ does. In Section 2.2.3, the realized competing event type X_T naturally arose as the state occupied by the process at event time T. This process point of view becomes indispensable when keeping track of an individual's course through multistate models as depicted in Figures 2.6 and 2.7. In all these models, we

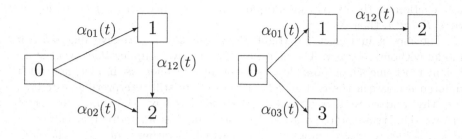

Fig. 2.6. Illness-death models without recovery.

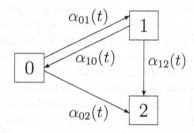

Fig. 2.7. Illness-death model with recovery.

write X_t for the state occupied by the individual at time t.

The model in Figure 2.6 (left) is a so-called illness-death model without recovery. The name of the model stems from the fact that in medical applications state 0 is often interpreted as 'healthy', state 1 is interpreted as 'diseased', and state 2 as 'dead'. The model is 'without recovery', because 'diseased' → 'healthy' transitions are not modelled. An individual may either start in the 'healthy' state or in the 'diseased' state (i.e., $X_0 \in \{0, 1\}$).

An individual that starts in the 'healthy' state will have either one or two event times. Say, T is the time the individual leaves its initial state 0. Then $X_t = 0$ for all $t \in [0, T)$ and $X_T \in \{1, 2\}$. If the individual makes a 'healthy' → 'dead' transition (i.e., a 0 → 2 transition and $X_T = 2$), there will be no further events for this individual. However, if the individual moves into the 'diseased' state 1 at time T (i.e., $X_T = 1$), there will be a future event time $\tilde{T}$, say, at which the individual moves from 'diseased' to 'dead', $X_{\tilde{T}} = 2$. An individual that starts in the 'diseased' state has only one event time, at which a 1 → 2 transition is made. Note again that the individual is in the 'target state' of a transition at the event time in question. If we have a $l \to j$ transition at time T, $l \neq j$, then $X_T = j$.

The model in Figure 2.6 (right) is a so-called progressive illness-death model without recovery. The attribute 'progressive' implies that every state has at most one single possible transition into it. There is, in fact, not much difference between the two models of Figure 2.6, still interpreting state 0 as 'healthy' and state 1 as 'diseased'. Both models have the same number of arrows (i.e., transition types), and in both models an individual enters an absorbing state at the time of 'death', which is either state 2 in the non-progressive model or state 2 or 3 in the progressive model. This means that states 2 and 3 of the progressive model have the same interpretation 'death'. The advantage of the progressive model is one of coding: by simply looking at the state entered at time of 'death', one is able to tell whether the individual was 'healthy' or 'diseased' just prior to 'death'. In the non-progressive model, one would also have to look at the state occupied just prior to 'death'.

The model in Figure 2.7 is called an illness-death model with recovery. The difference compared to the models in Figure 2.6 is that now 'diseased' $\to$ 'healthy' transitions (i.e., recoveries) are also possible. Already in the model without recovery, we saw that the number of an individual's event time can be random. However, there were at most two subsequent events. In a model with recovery, there is, at least theoretically, no such maximum number. When actually collecting data, practical restrictions may impose a maximum number. Still, the number of an individual's event times would be random, which further stresses the process way of writing things. As in Section 2.1, an individual's course through model Figure 2.7 is still conveniently written as

$$[0, \infty) \ni t \mapsto X_t,$$

$X_t \in \{0, 1, 2\}$.

Obviously, multistate models can be quite complex. In principle, any finite number of states is admissible, and there may be transitions in both directions between every pair of states. E.g., if states 0 and 1 correspond to two operational levels of a machine, say 'on' and 'off', and state 2 corresponds to 'malfunctioning', the machine may be repaired, such that transitions out of state 2 are also possible. This model specification (i.e., the state space and the possible transitions types between the states) will be directly reflected when using the R packages mvna, etm, and mstate. We also refer to the excellent textbook by Hougaard (2000) who gives a comprehensive account of the different types of multistate models.

Multistate models as a series of nested competing risks experiments

As stated earlier, the multistate models that we consider are realized as successive nested competing risks experiments. For illustration, consider the illness-death model with recovery of Figure 2.7. An individual either starts in state 0

or in state 1 at time origin 0. Consider an individual that starts in state 0;
$X_0 = 0$. The course of this individual through the multistate model is realized
as follows.

1. Being in state 0, the individual is exposed to 'cause-specific' hazards $\alpha_{01}(t)$
   and $\alpha_{02}(t)$. As explained at the end of Section 2.2.3, the individual's waiting time until an event, the time until moving out of state 0, is determined
   by the 'all-cause' hazard $\alpha_{01}(t) + \alpha_{02}(t)$. The state entered at the time
   of transition is determined by a binomial experiment and the relative
   magnitude of the 'cause-specific' hazards $\alpha_{01}(t)$ and $\alpha_{02}(t)$ at the time of
   transition.
2. The next step depends on the state entered at the first transition time.
   a) If the absorbing state 2 has been entered, there will be no further
      transition.
   b) If state 1 has been entered, a new competing risks experiment is carried
      out using the current values of the 'cause-specific' hazards $\alpha_{10}(t)$ and
      $\alpha_{12}(t)$. Otherwise, the experiment runs analogously to step 1.
3. The next step depends on the state entered at the second transition time.
   If the absorbing state has been entered, there will be no further transition.
   If state 0 has been entered, a further competing risks experiment will be
   carried out.

This series of competing risks experiments will be carried out until absorption.
If there is no absorbing state in the model (i.e., backward transitions are
feasible out of every state), we will need to keep track of the series of competing
risks experiments until observation ends.

As with competing risks, this simulation point of view towards multistate
models shows up again and again in the main part of the book. See, in particular, Chapter 8 for an in-depth treatment. A more formal justification of the
above algorithm is given in Section 4.4 of Gill and Johansen (1990); see also
Theorem II.6.7 of Andersen et al. (1993).

Transition probabilities and transitions hazards of a time-inhomogeneous Markov multistate process

So far, our treatment of multistate models has been a bit lax in that the
transitions hazards $\alpha_{lj}(t)$, $l \neq j$, which we indicated at the $l \rightarrow j$ arrows in the
multistate figures, have been treated as cause-specific hazards of a competing
risks model in the above algorithm. Although conceptually correct, a slightly
more precise definition is desirable. We also wish to estimate the cumulative
transition hazards and derive probability estimates, and we need to state in
a more precise manner that multistate models that are realized as successive
nested competing risks experiments are time-inhomogeneous Markov.

The Markov property is a key assumption for the estimation techniques
discussed below to work with data subject to independent left-truncation and
independent right-censoring, respectively. In essence, the Markov property,

which is given a precise form in Equation (2.27) below, means that the future course of an individual depends on the past only via the current time and the state currently occupied by the individual. E.g., the future development of a 'diseased' individual at time t in the models of Figure 2.6 depends on the past only through the time elapsed since time origin (i.e., t and the fact that the individual is currently 'diseased') but not on the time span the individual has already been ill. The analysis of non-Markov models is a quite active research field, on which we briefly comment in Chapter 12.

We begin by defining the matrix of transition probabilities of a Markov process $(X_t)_{t \geq 0}$ with state space $\{0, 1, 2, \ldots, J\}$ as

$$\mathbf{P}(s, t) := (\mathrm{P}_{lj}(s, t))_{l,j}, \, l, j \in \{0, 1, 2, \ldots, J\}, \qquad (2.26)$$

with transition probabilities

$$\mathrm{P}_{lj}(s, t) := \mathrm{P}(X_t = j \mid X_s = l) = \mathrm{P}(X_t = j \mid X_s = l, \mathrm{Past}), s \leq t. \qquad (2.27)$$

The Markov property $\mathrm{P}(X_t = j \mid X_s = l) = \mathrm{P}(X_t = j \mid X_s = l, \mathrm{Past})$ intuitively states that past and future of the process are independent given the present at time s. We also note that the Markov process $(X_t)_{t \geq 0}$ is said to be time-inhomogeneous, because the transition probabilities (2.27) depend on the actual time interval $[s, t]$. In contrast, a homogeneous process makes the more restrictive assumption that these probabilities are identical whenever the length of the time interval $d = t - s$ is. The transition probabilities of a homogeneous Markov process only depend on the length of the time interval, but not the interval itself. Readers should note that sometimes homogeneous Markov processes are simply called 'Markov processes', dropping the attribute 'homogeneous'.

Analogous to Definition (2.23) of the cause-specific hazards, we now define the transition hazards of the Markov process

$$\alpha_{lj}(t) \cdot dt := \mathrm{P}(X_{(t+dt)-} = j \mid X_{t-} = l), \, l, j = 0, \ldots, J, \, l \neq j. \qquad (2.28)$$

Note that the Markov property implies that conditioning on $X_{t-} = l$ is tantamount to conditioning on the entire past of the process before t. In words, $\alpha_{lj}(t) \cdot dt$ is the probability of making an $l \to j$ transition in the very small time interval dt. Intuitively, dt will be so small that the transition occurs directly from l to j (i.e., without visiting another state in between). Thus, we can think of $\alpha_{lj}(t)$ as momentary forces of transition between states l and j. Formally, we also define

$$\alpha_{ll}(t) = - \sum_{j=0, j \neq l}^{J} \alpha_{lj}(t), \, l = 0, \ldots, J. \qquad (2.29)$$

This definition is justified following Equation (2.31) below.

At the beginning of this chapter, we claimed that transition hazards are an intuitively important concept. And, in fact, the $\alpha_{lj}(t)$s of (2.28) are well suited to contrast time-inhomogeneous Markov models from homogeneous models and from non-Markov models, respectively.

In a homogeneous Markov model, $\mathbf{P}(s,t)$ only depends on $t - s$, but not on the actual time interval. As a consequence, $\alpha_{lj}(t) = \alpha_{lj}(0)$ for all t: a homogeneous model is a parametric model with constant transition hazards. In contrast, the transition hazards (2.28) can essentially be any integrable nonnegative function. Hence, assuming $(X_t)_{t \geq 0}$ to be time-inhomogeneous Markov provides for a much larger nonparametric model.

The restriction implied by the Markov assumption is also well illustrated in terms of the transition hazards. In the illness-death models of Figure 2.6, the 'illness' $\rightarrow$ 'death'-hazard is $\alpha_{12}(t)$. It depends on the transition type 'illness' $\rightarrow$ 'death' and on the current time t since time origin 0. However, $\alpha_{12}(t)$ does not depend on the entry time $\tilde{t}$, say, into the 'illness'-state 1, $\tilde{t} < t$. In a non-Markov model, the transition hazard would be $\alpha_{12}(\tilde{t}, t)$, which would potentially be different for fixed t but different times $\tilde{t}$ of falling ill.

Nelson-Aalen estimator

As our Definition (2.28) of the transition hazards has been analogous to Definition (2.23) of the cause-specific hazards, it should not come as a surprise that we may estimate the cumulative transition hazards $A_{lj}(t) = \int_0^t \alpha_{lj}(u) \, du$ in a manner similar to the cause-specific Nelson-Aalen estimators (2.25). The appropriate Nelson-Aalen estimators are

$$\widehat{A}_{lj}(t) = \sum_{k=1}^{K} \frac{\text{number of observed } l \rightarrow j \text{ transitions at } t_k}{\text{number of individuals at risk in state } l \text{ just prior to } t_k}, \quad (2.30)$$

$l, j = 0, \dots, J$, $l \neq j$, where the summation is over all event times t_k, which are less than or equal to t. As with the cause-specific Nelson-Aalen estimators, we stress a couple of important facts which have been alluded to earlier: the numerator in (2.30) represents increments of a transition-specific counting process. And the risk set in (2.30) includes all individuals who have entered state l before time t_k and who have not yet moved out of state l again or have been censored.

This has three important implications. First, as with the cause-specific Nelson-Aalen estimators, we may code computation of $\widehat{A}_{lj}(t)$ via coding $l \rightarrow \tilde{j}$ transitions, $\tilde{j} \neq j$, as censoring events and only count type $l \rightarrow j$ transitions as 'actual events'. Second, an individual only contributes to the risk set in state l after entry into the state; movements within a multistate model generate 'internal' left-truncation as explained towards the end of Section 2.2.2. An analysis of $\widehat{A}_{lj}(t)$ must be coded accordingly. Third, every individual in state l and under observation contributes to the risk set alike; there is no further accounting for the individual's entry time into state l. This is a consequence of the Markov assumption. Each of these implications will be directly reflected in R coding.

In fact, one may adopt the view that these three implications are consequences of multistate models being realized as a series of competing risks experiments. It then is via the methodology of Sections 2.2.1 and 2.2.2 that we may analyse the transition hazards of, first, competing risks, and, next, multistate models.

Product integration and the Aalen-Johansen estimator

Finally, we wish to estimate the matrix of transition probabilities $\mathbf{P}(s,t)$. In the simple survival set-up of Section 2.1, we found that we may compute the survival probability as a deterministic function, namely product integration, of the cumulative survival hazard. Replacing the true cumulative hazard by its Nelson-Aalen estimator resulted in the Kaplan-Meier estimator of the survival function. An analogous approach works for estimating $\mathbf{P}(s,t)$.

Analogous to $\mathbf{P}$, we write

$$\mathbf{A}(t) := (A_{lj}(t))_{l,j}, \ l,j \in \{0,1,2,\ldots,J\} \tag{2.31}$$

for the matrix of cumulative transition hazards $A_{lj}(t) = \int_0^t \alpha_{lj}(u)\,\mathrm{d}u$. The aim is to show that $\mathbf{P}(s,t)$ can be computed as a continuous matrix-valued product over terms

$$\mathbf{I} + \mathrm{d}\mathbf{A}(u),$$

where u ranges from s to t, where we have written $\mathbf{I}$ for the $(J+1) \times (J+1)$ identity matrix, and where $\mathrm{d}\mathbf{A}(u)$ is defined element wise as

$$\mathrm{d}\,(A_{lj}(u))_{l,j} = (\alpha_{lj}(u))_{l,j}\,\mathrm{d}u,$$

$l,j \in \{0,1,2,\ldots,J\}$. This idea obviously parallels that of Equations (2.5)–(2.7). Also recall from (2.29) that $\mathrm{d}A_{ll}(u) = -\sum_{j=0,j\neq l}^{J} \mathrm{d}A_{lj}(t)$, such that

$$1 - \mathrm{d}A_{ll}(u) = 1 - \mathrm{P}(X_{(t+\mathrm{d}t)-} \neq l \,|\, X_{t-} = l) = \mathrm{P}(X_{(t+\mathrm{d}t)-} = l \,|\, X_{t-} = l),$$

which explains Definition (2.29), or equivalently why we have to consider a product over terms $\mathbf{I} + \mathrm{d}\mathbf{A}(u)$.

Now consider a time v, $s < v < t$. The Markov property implies that

$$\mathbf{P}(s,t) = \mathbf{P}(s,v) \cdot \mathbf{P}(v,t). \tag{2.32}$$

In order to see that (2.32) holds, consider the (l,j)th entry of $\mathbf{P}(s,t)$:

$$\mathrm{P}(X_t = j \,|\, X_s = l) = \sum_{\tilde{j}=0}^{J} \mathrm{P}(X_v = \tilde{j} \,|\, X_s = l) \cdot \mathrm{P}(X_t = j \,|\, X_v = \tilde{j}, X_s = l)$$

$$= \sum_{\tilde{j}=0}^{J} \mathrm{P}(X_v = \tilde{j} \,|\, X_s = l) \cdot \mathrm{P}(X_t = j \,|\, X_v = \tilde{j}),$$

where the last equation holds because of the Markov property, and the right hand side equals the (l, j)th entry of the right hand side of (2.32).

Next, assume that v is close to t such that an approximation such as in Equation (2.7) holds,

$$\sum_{\tilde{j}=0}^{J} P(X_v = \tilde{j} \mid X_s = l) \cdot (\mathbf{1}(\tilde{j} = j) + \Delta A_{\tilde{j}j}(t)),$$

where we have written $\Delta A_{\tilde{j}j}(t)$ for $A_{\tilde{j}j}(t) - A_{\tilde{j}j}(v)$. Doing this recursively for a partition $s = t_0 < t_1 < t_2 < \ldots < t_{K-1} < t_K = t$ of the time interval $[s, t]$, we get the approximation

$$\mathbf{P}(s, t) \approx \prod_{k=1}^{K} (\mathbf{I} + \Delta \mathbf{A}(t_k)), \tag{2.33}$$

where the (l, j)th element of $\Delta \mathbf{A}(t_k)$ is $A_{lj}(t_k) - A_{lj}(t_{k-1})$. Computing the approximation on the right hand side of (2.33) for ever finer partitions of $[s, t]$ approaches a limit, the matrix-valued product integral $\prod_{u \in (s,t]} (\mathbf{I} + d\mathbf{A}(u))$. The product integral equals the matrix of transition probabilities,

$$\mathbf{P}(s, t) = \prod_{u \in (s,t]} (\mathbf{I} + d\mathbf{A}(u)). \tag{2.34}$$

An estimator of $\mathbf{P}(s, t)$ is now naturally derived by replacing $\mathbf{A}(u)$ with the matrix $\widehat{\mathbf{A}}(u)$ of Nelson-Aalen estimators with the (l, j)th entry $\widehat{A}_{lj}(u)$ as in Equation (2.30) for $l \neq j$ and $\widehat{A}_{ll}(u) := -\sum_{j, j \neq l} \widehat{A}_{lj}(u)$. We also define $d\widehat{\mathbf{A}}(u)$ as the matrix with entries $\widehat{A}_{lj}(u) - \widehat{A}_{lj}(u-)$ (i.e., the increment of the Nelson-Aalen estimators at time u). This results in the Aalen-Johansen estimator (Aalen and Johansen, 1978),

$$\widehat{\mathbf{P}}(s, t) = \prod_{u \in (s,t]} \left(\mathbf{I} + d\widehat{\mathbf{A}}(u) \right), \tag{2.35}$$

which is an ordinary, finite matrix product over all event times u in $(s, t]$ and matrices $\mathbf{I} + d\widehat{\mathbf{A}}(u)$. The Aalen-Johansen estimator is often also called the empirical transition matrix. The estimator and 'empirical' product integration are implemented in the R package etm.

We finally note that checking approximation (2.33) in R as wo did for the simple survival situation in Section 2.1 is not straightforward. The reason is that closed formulae for $\mathbf{P}(s, t)$ only exist for some special, practically important multistate models; see Section 9.1. In fact, approximation (2.33) provides a numerical tool to compute $\mathbf{P}(s, t)$ in the absence of closed formulae.

2.3 Approximate inference in practice based on large sample results

As with standard survival data, statistical inference for competing risks and multistate models is typically of a nonparametric kind. Asymptotic results are used for approximate inference in practice. E.g., the transition-specific Nelson-Aalen estimator $\widehat{A}_{lj}(t)$ as defined in (2.30) is approximately unbiased in the sense that $\widehat{A}_{lj}(t)$ converges in probability to the true quantity $A_{lj}(t)$. Properly standardized, the distribution of the estimator approaches a normal distribution,

$$\sqrt{n}\left(\widehat{A}_{lj}(t) - A_{lj}(t)\right) \to N(0, \sigma_{lj}^2(t)), \tag{2.36}$$

where n is the number of individuals under study. An (again asymptotically/approximately unbiased) estimator of the asymptotic variance $\sigma_{lj}^2(t)$ is $n \cdot \widehat{\sigma}_{lj}^2(t)$, where

$$\widehat{\sigma}_{lj}^2(t) = \sum_{k=1}^{K} \frac{\text{number of observed } l \to j \text{ transitions at } t_k}{(\text{number of individuals at risk in state } l \text{ just prior to } t_k)^2},$$

where the summation is over all event times t_k, which are less than or equal to t as in (2.30); see also Section 4.1. This can be used, e.g., to construct an approximate 95% confidence interval

$$\widehat{A}_{lj}(t) \pm \widehat{\sigma}_{lj}(t) \cdot 1.96,$$

where $1.96 \approx$ qnorm(0.975), i.e., the 0.975 quantile of the standard normal distribution. (We note, however, that a log-transformed confidence interval should be preferred in small samples; see (4.10).)

A common feature of these approximate procedures is, loosely speaking, that they only hold on the 'observable time interval'. For continuous event times and continuous censoring times, the 'observable time interval' is restricted by the upper limit of the joint support of event time and censoring time. In practice, the 'observable time interval' is considered to be restricted by the largest uncensored event time. Analogous considerations are needed for left-truncated data and for more complex multistate models. A detailed discussion can be found in Examples IV.1.6–IV.1.9 in Andersen et al. (1993). A sufficient condition for the approximate procedures to work is that there is a positive probability of being at risk in a transient state of the multistate model under consideration (Andersen et al., 1993, Equation (4.1.16)). Transient states are those states out of which a transition is possible. E.g., in the competing risks model of Figure 2.5, only the initial state 0 is transient, but in the illness-death models of Figures 2.6 and 2.7, both states 0 and 1 are transient. An individual is said to be at risk in a transient state l just prior to time t, if the individual is in state l and under observation at $t-$. Only such an individual may be observed to make a transition out of state l at time t.

Requiring a positive probability of a non-empty risk set has different implications depending on which quantity is being estimated. If the aim is to estimate the cumulative transition hazard between states l and j of a certain multistate model, $l \neq j$, the requirement only affects the risk set Y_l. Estimation of probabilities, however, in general depends on the estimation of all cumulative hazards because of relations (2.34) and (2.35).

For illustration, we briefly comment on a basic implication in the simple competing risks model. Often, there is interest in the failure type probabilities $P(X_T = j)$, where j is one of the competing event states $1, 2, \ldots, J$ as in Figure 2.5. E.g., Mackenbach et al. (1999) investigated such 'prevalences' of causes of death in the Netherlands. $P(X_T = j)$ is the limit of the so-called cumulative incidence function,

$$P(X_T = j) = \lim_{t \to \infty} P(T \leq t, X_T = j).$$

Estimation of $P(T \leq t, X_T = j)$ is a special application of the Aalen-Johansen estimator (2.35). Estimation of its limit $P(X_T = j)$ is simple, if the data are complete. The Aalen-Johansen estimator evaluated at the largest time point then simply equals the so-called 'crude rates', the number of type j events (at any time), divided by the sample size n. However, $P(X_T = j)$ will not be nonparametrically estimable with most right-censored data. The reason for this is that censoring typically restricts the 'observable time interval' such that one will not be able to observe the limit of the cumulative incidence function. See also our discussion following (4.21).

It is worthwhile to note that approximate unbiasedness and approximate normality hold *uniformly* on what we have loosely called an 'observable time interval'. For weak convergence, this requires a theory of convergence of probability measures on a space of functions rather than the well-known concept of weak convergence of distribution functions. This functional point of view is, e.g., relevant when moving from the Nelson-Aalen estimator $\widehat{\mathbf{A}}$ to the Aalen-Johansen estimator $\widehat{\mathbf{P}}(s,t)$; see (2.35). This is so, because $\widehat{\mathbf{P}}(s,t)$ is a function of *all* previous Nelson-Aalen estimates between s and t, i.e., all $\widehat{\mathbf{A}}(u)$, $u \in (s,t]$.

The mathematics of such a convergence theory are formidable and beyond the technical level of this book. We are content with the fact that asymptotic unbiasedness and asymptotic normality can be formulated rigorously and sufficiently general for the applications in this book. The generally interested reader is referred to the excellent books by Billingsley (1968), Andersen et al. (1993), and van der Vaart and Wellner (1996). In particular, Billingsley's introduction gives a very readable account of why a functional point of view is useful. On the other hand, his Section 3.18 concisely explains why obtaining *uniform* results is difficult. These difficulties have been solved using the modern theory of empirical processes, of which van der Vaart and Wellner give a definite account. Finally, Andersen et al. give a dense but thorough description of asymptotic theory for event history analysis; see, in particular, their Section II.8.

We must, however, still mention the functional delta method as an important tool from the theory of empirical processes. The usual delta method starts with standard, i.e., pointwise convergence as in (2.36). Considering a differentiable transformation $\phi : \mathbb{R} \to \mathbb{R}$ with derivative ϕ', the delta method implies that

$$\sqrt{n}\left(\phi(\widehat{A}_{lj})(t) - \phi(A_{lj})(t)\right) \to \phi'(A_{lj}(t))N(0, \sigma_{lj}^2(t)), \qquad (2.37)$$

and that the left-hand side of (2.37) is asymptotically equivalent to $\phi'(A_{lj}(t)) \cdot \sqrt{n}\left(\widehat{A}_{lj}(t) - A_{lj}(t)\right)$, i.e., the difference of the asymptotically equivalent terms converges to zero in probability. We use the ordinary delta method for obtaining pointwise confidence intervals of the Nelson-Aalen estimator based on a log-transformation; see (4.10). A generalization of the delta method to p vectors and transformations $\phi : \mathbb{R}^p \to \mathbb{R}^q$ is immediately available, but what is really needed is a *functional* delta method. This does exist (Gill, 1989), but is again beyond the technical level of this book. We note, however, that the functional delta method works for product integration as in Equations (2.34) and (2.35). This further emphasizes the key roles played by both the Nelson-Aalen estimator and the product integral. We mention two further important consequences. The functional delta method preserves asymptotic normality and it justifies using bootstrap resampling as discussed in Appendix A; see Gill (1989) and van der Vaart and Wellner (1996). This is helpful in situations where variance estimators are analytically hardly tractable. The variance may then be estimated based on the bootstrap and confidence intervals may again be constructed based on approximate normality.

2.4 Exercises

1. Show that an event time T with hazard $\alpha(t)$ has distribution function $P(T \leq t) = 1 - \exp(-\int_0^t \alpha(u)\, du)$.
2. Write a function A.gompertz for the cumulative hazard when the survival time distribution follows a Gompertz distribution with shape parameter $\lambda = 1$ and scale parameter $\gamma = 2$.
   Under the Gompertz distribution, the hazard is

$$\alpha(t) = \lambda \exp(t/\gamma).$$

   Using the prodint function from Section 2.1.1, approximate $S(1) = P(T > 1)$ and compare it to the true value.
3. Simulate 100 individuals with survival times following a Gompertz distribution with parameters as in Exercise 2. Also simulate independent censoring times following a uniform distribution in order to obtain approximately 30% of censored observations. A function to generate Gompertz random variables can be found in the R package eha.

Estimate $S(1)$ both using `prodint` and the `survfit` function.

Using the output of `survfit`, compute the Nelson-Aalen estimator of the cumulative hazard and check whether it is close to the true cumulative hazard function.

4. Plot the counting process of observed events and its compensator as in Figure 2.3. Check that the compensator is only almost a smooth line by displaying some 'edges'.

5. Redo the analysis of Exercise 3, but this time with at least 50% censoring.

6. Reuse the simulated data set from Exercise 3 and additionally simulate independent left-truncation times which follow a Weibull distribution. Choose the parameters such that approximately 70% of the simulated individuals are actually included in the study. Check that the Nelson-Aalen estimator 'works' in the presence of left-truncation and right-censoring.

7. *Competing risks*: Simulate competing risks data for 200 individuals with constant cause-specific hazards $\alpha_{01}(t) = 0.5$ and $\alpha_{02}(t) = 0.9$. Independent right-censoring times follow a uniform distribution with parameter chosen to give approximately 20% of censored observations.

   Compute the cause-specific Nelson-Aalen estimators.

8. Definition (2.22) of the cause-specific hazards implies that

$$P(T \le t, X_T = 1) = \int_0^t P(T \ge u-)\alpha_{01}(u)\,du.$$

   Show that

$$P(T \le t, X_T = 1) \le 1 - \exp(-\int_0^t \alpha_{01}(u)\,du).$$

   One minus the right hand side of the previous equation is sometimes called the 'cause-specific survivor function'. The right hand side of the equation can be estimated using one minus a Kaplan-Meier-type estimator, but it lacks a proper probability interpretation. Check that this estimator overestimates $P(T \le t, X_T = 1)$ using the simulated competing risks data.

9. *Multistate models*: Simulate data from an illness-death model without recovery. All individuals are assumed to start in an initial state 0, 'healthy'. Hazards out of the initial state are as in Exercise 7. For individuals who reach state 1, simulate new event times $\tilde{T}$ with constant hazard $\alpha_{12}(t) = 0.8$. $T + \tilde{T}$ will then be the time of entry into state 2, 'death', for individuals who have moved through the 'illness' state 1.

   Estimate the cumulative transition hazards for the following scenarios.

   a) Complete data.

   b) Randomly right-censored data: Draw uniformly distributed censoring times C such that approximately 20% of the observations are censored in the initial state.

    c) State-dependent censoring: Assume that individuals who are *observed* to move into state 1 are subject to censoring times $\tilde{C}$ which follow a uniform distribution that is different from the distribution of C.

    d) Repeat the previous analyses, but additionally introduce random left-truncation, with left-truncation times stemming from a gamma distribution with parameters chosen to let approximately 90% of the individuals enter the study.

10. *Time-inhomogeneous Markov property*: Show that a competing risks process fulfills the Markov property. When is an illness-death model without recovery Markov?

Part II

Competing risks

3

Multistate modelling of competing risks

Competing risks models analyse the time until some first event *and* the event type that occurs at that time. In contrast, standard survival analysis considers the time until some first event only. Examples include disease-free survival and length of hospital stay. Disease-free survival is observed either at the time the disease in question is diagnosed or at the time of death without prior disease. Length of hospital stay ends with either discharge alive or hospital death. In the latter example, survival models consider length of stay with a combined endpoint discharge alive/hospital death. Competing risks also model the endpoint type. Competing risks do not model subsequent events such as death after hospital discharge. To do this, more complex multistate models are needed, which is the topic of the multistate part of this book.

The present Chapter 3 introduces competing risks as a multistate model, key quantities such as the cause-specific hazards and the cumulative incidence functions, and simulation of competing risks data. Some readers may be familiar with a different approach towards competing risks, namely the latent failure time model. This model, its inherent difficulties, and the advantages of the multistate approach taken in this book are briefly illustrated in Section 3.3.

Although estimation from competing risks data is more prominent in everyday work, we have chosen also to present competing risks simulation, because it is extremely helpful to understand the key concepts. Simulation can be viewed as the practical aspect of the probabilistic task of constructing a competing risks process. Besides using real data, we also simulate competing risks data in R and analyse them in order to see that the proposed methodology works. I.e., the methodology is able to recover the underlying data-generating mechanism even in the presence of right-censoring and left-truncation. Simulation is also useful to understand the *role* that the estimated quantities play. This is of vital importance, because interpretation of results often is a major challenge in the analysis of competing risks data. Finally, 'playing around' with simulated data further helps to understand the key concepts.

As competing risks generalize survival analysis from a single combined endpoint to multiple first event types, our treatment of competing risks also

covers key survival concepts and their analysis in R. An elementary introduction to single endpoint survival analysis in R is given by Dalgaard (2002,
Chapter 12).

3.1 The competing risks multistate model

Figure 3.1 depicts the standard competing risks multistate model. Initially,

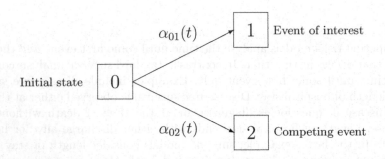

Fig. 3.1. Competing risks multistate model with cause-specific hazards $\alpha_{0j}(t)$, $j =$
$1, 2$.

every individual is in the initial state 0 at time origin. The individual stays
in this state until occurrence of any first event. Usually, there is one event of
interest, modelled by transitions into state 1, and all other first event types
are subsumed into the competing event state 2. E.g., in the case of hospital
stay data, hospital mortality modelled by state 1 may be of special interest.
The competing event state 2 then has the interpretation 'discharge alive'.
However, this competing event state does not further distinguish whether the
patient returns home, is readmitted to a different hospital, or enters a nursing
home. The competing risks techniques of this book allow for more than two
competing event states; see Section 7.1. As the model of Figure 3.1 is the
single most important competing risks model, we focus on two competing
event states for ease of presentation. The techniques then easily generalize to
more than two competing risks as explained in Section 7.1. An R description
of the competing risks multistate model in Figure 3.1 is given in the data
analysis of Chapter 4.

Recall from Chapter 2 that we need to keep track of an individual's
movements over the course of time. The competing risks process $(X_t)_{t\geq0}$
of Figure 3.1 denotes the state an individual is in for every point in time,
$X_t \in \{0, 1, 2\}$. Every individual starts in the initial state 0 at time origin 0,

$$P(X_0 = 0) = 1. \tag{3.1}$$

An individual stays in state 0 (i.e., $X_t = 0$) as long as neither competing event 1 nor 2 has occurred. The individual moves to state 1 if the event of interest occurs. Likewise, the individual moves to state 2 if the other competing event occurs first. The competing risks process moves out of the initial state 0 at time T,

$$T := \inf\{t > 0 \,|\, X_t \neq 0\}. \tag{3.2}$$

T is often called survival time or failure time; it can be thought of as a waiting time in the initial state 0. In many applications, every individual will leave the initial state at some point in time, i.e., $P(T \in (0, \infty)) = 1$. Hence, we typically think of T as a nonnegative, real-valued event time, although $T < \infty$ is usually not required from a mathematical point of view.

Recall from Chapter 2 that multistate processes are right-continuous (cf. also Figure 2.2). The competing risks process is either in state 1 or in state 2 at time T. The type of the first event, often called cause of failure, therefore is

$$X_T \in \{1, 2\}, \tag{3.3}$$

the state the process enters at time T.

Observation of the competing risks process $(X_t)_{t \geq 0}$ will in general be subject to right-censoring and/or left-truncation as explained in Chapter 2. If observation of the process is subject to a right-censoring time C only, the observed data will be $(T \wedge C, \mathbf{1}(T \leq C) \cdot X_T)$. Note that the status indicator

$$\mathbf{1}(T \leq C) \cdot X_T \in \{0, 1, 2\} \tag{3.4}$$

equals 0, if the observation was censored. Note that in Chapter 4, we find that from a *software coding* perspective it may be advantageous to code censoring events by a value different from 0; we often use 'cens' as a censoring code. In case of a censoring event, observation stopped while the individual was still in the initial state 0. Otherwise, failure time and cause of failure have been observed, and the status indicator equals X_T. If we also have a left-truncation time L, the observed data will be $([L, T \wedge C], \mathbf{1}(T \leq C) \cdot X_T)$. In Chapter 4 on nonparametric estimation from competing risks data, we consider n independent replicates of (T, X_T), subject to independent right-censoring and left-truncation as explained in Section 2.2.2.

Key quantities in competing risks are the cause-specific hazards $\alpha_{0j}(t)$, $j = 1, 2$,

$$\alpha_{0j}(t)dt := P(T \in dt, X_T = j \,|\, T \geq t), \, j = 1, 2. \tag{3.5}$$

Recall from Chapter 2 that we write dt both for the length of the infinitesimal interval $[t, t+dt)$ and the interval itself. We also write $A_{0j}(t)$ for the cumulative cause-specific hazards

$$A_{0j}(t) := \int_0^t \alpha_{0j}(u)du, \, j = 1, 2. \tag{3.6}$$

The cause-specific hazards can be thought of as momentary forces of transition, moving along the arrows in Figure 3.1. It is crucial to any competing risks analysis that both cause-specific hazards completely determine the stochastic behaviour of the competing risks process (cf. Sections 2.2.3–2.2.4). The α_{0j}s sum up to the all-cause hazard:

$$\alpha_{0\cdot}(t)\mathrm{d}t := p(T \in \mathrm{d}t \mid T \geq t) \tag{3.7}$$

$$= (\alpha_{01}(t) + \alpha_{02}(t))\, \mathrm{d}t. \tag{3.8}$$

This result is a consequence of the usual additivity of probabilities. We also write

$$A_{0\cdot}(t) := \int_0^t \alpha_{0\cdot}(u)\mathrm{d}u = A_{01}(t) + A_{02}(t) \tag{3.9}$$

for the cumulative all-cause hazard.

The survival function of the waiting time T in the initial state 0 is:

$$P(T > t) = \exp(-\int_0^t \alpha_{0\cdot}(u)\, \mathrm{d}u). \tag{3.10}$$

The survival function $P(T > t)$ is often denoted $S(t)$, cf. Section 2.1. It is a function of both α_{0j}s, because $\alpha_{0\cdot}(t) = \alpha_{01}(t) + \alpha_{02}(t)$.

Often, interest in competing risks focuses on the cumulative incidence function, i.e., the expected proportion of individuals experiencing a certain competing event over the course of time:

$$P(T \leq t, X_T = j) = \int_0^t P(T > u-)\,\alpha_{0j}(u)\, \mathrm{d}u, \ j = 1, 2, \tag{3.11}$$

where we write $P(T > u-)$ for the value of the survival function just prior to u. We note that the cumulative incidence function is a function of both α_{0j}s through $P(T > u-)$.

There is an intuitive interpretation of the right hand side of Equation (3.11), which shows up again when we consider nonparametric estimation in Chapter 4. One integrates or, loosely speaking, sums up over 'infinitesimal probabilities' of making the $0 \to j$ transition exactly at time u. $P(T > u-)$ is the probability of still being in the initial state 0 before u; this probability is multiplied with the 'infinitesimal transition probability' $\alpha_{0j}(u)\, \mathrm{d}u$ to actually make the $0 \to j$ transition at time u conditional on still being in the initial state before u (cf. Equation (3.5)).

The right hand side of (3.11) is the solution of a product integral as in Section 2.2.4. More precisely, $P(T \leq t, X_T = 1)$ is the entry $(1,2)$ of the solution of

$$\prod_{u \in (0,t]} \left\{ \begin{pmatrix} 1 & 0 & 0 \\ 0 & 1 & 0 \\ 0 & 0 & 1 \end{pmatrix} + \mathrm{d} \begin{pmatrix} -A_{0\cdot}(u) & A_{01}(u) & A_{02}(u) \\ 0 & 0 & 0 \\ 0 & 0 & 0 \end{pmatrix} \right\};$$

see Equation (2.34). Entry $(1,3)$ is $P(T \leq t, X_T = 2)$, and the survival function $P(T > t)$ is in entry $(1,1)$. Entries in rows 2 and 3 are as rows 2 and 3 of the 3×3 identity matrix, reflecting that competing risks do not model events after the first event. We meet this structure again when estimating the cumulative incidence functions (cf. Table 4.1).

We finally note that the cumulative incidence function for the event state 1 of interest, $P(T \leq t, X_T = 1)$, and for the competing event state 2, $P(T \leq t, X_T = 2)$, add up to the all-cause distribution function,

$$P(T \leq t, X_T = 1) + P(T \leq t, X_T = 2) = P(T \leq t), \qquad (3.12)$$

and that the cumulative incidence functions approach $P(X_T = j)$, $j = 1, 2$ as time t increases,

$$\lim_{t \to \infty} P(T \leq t, X_T = j) = P(X_T = j), \, j = 1, 2, \qquad (3.13)$$

assuming T to be a finite time.

We consider nonparametric estimation of the cumulative cause-specific hazards, the all-cause survival function, and the cumulative incidence functions in Chapter 4. But before we do so, we first consider how competing risks data can be simulated cause-specific hazard-based in Section 3.2.

3.2 Simulating competing risks data

In Section 3.1, we claimed that the cause-specific hazards completely determine the stochastic behaviour of the competing risks process. The aim of this section is to generate competing risks data from a given pair of cause-specific hazards $\alpha_{01}(t)$ and $\alpha_{02}(t)$. The simulation algorithm is also the key building block for simulating more complex multistate data. The aim of our presentation of the simulation algorithm is to illustrate how a competing risks process is constructed. Understanding this construction helps to interpret competing risks analyses of real data. Later, in Chapter 4, we also analyse simulated data in order to see that the presented methodology works.

Recall from Section 3.1 that completely observed competing risks data are replicates of (T, X_T), where T is the failure time and X_T is the failure cause. Observation of (T, X_T) may be subject to a right-censoring variable C and to a left-truncation variable L. We generate competing risks data via the following algorithm.

1. Specify the cause specific hazards $\alpha_{01}(t)$ and $\alpha_{02}(t)$.
2. Simulate failure times T with all-cause hazard $\alpha_{0.}(t) = \alpha_{01}(t) + \alpha_{02}(t)$.
3. Run a binomial experiment for a simulated failure time T, which decides with probability $\alpha_{0j}(T)/(\alpha_{01}(T) + \alpha_{02}(T))$ on cause $j, j = 1, 2$.
4. Additionally generate right-censoring times C and left-truncation times L.

The censoring times C and, if desired, the truncation times L in Step 4 are often simulated as random variables independent of the competing risks process. This is the random censorship model and the random truncation model, respectively. A simple example of simulated competing risks data conditional on covariates is given in Section 5.2.2.

The competing risks specialty of the above algorithm is Step 3: Compared with standard survival simulation, we additionally have to generate a failure cause. This step also helps to understand what cause-specific hazards do. In Section 3.1, we suggested thinking of the α_{0j}s as momentary forces of transition, moving along the arrows in Figure 3.1. Figuratively speaking, the α_{0j}s are forces that pull an individual out of the initial state 0 and towards the competing risks states. Assuming that an individual fails at time $T = t$, the probability that the failure cause is 1 equals the proportion that the cause-specific hazard for failure 1 at time t contributes to the all-cause hazard $\alpha_{01}(t) + \alpha_{02}(t)$, which pulls the individual. More formally, given an individual fails at time $T = t$, the probability that failure is of type 1 is

$$P(X_T = 1 \,|\, T \in dt, T \geq t) = \frac{P(T \in dt, X_T = 1 \,|\, T \geq t)}{P(T \in dt \,|\, T \geq t)}$$

$$= \frac{\alpha_{01}(t)}{\alpha_{01}(t) + \alpha_{02}(t)}. \tag{3.14}$$

Generating replicates of T (and also C and L) can often be done in R with the help of convenience functions. If the all-cause hazard is constant (i.e., $\alpha_{0\cdot}(t) = \alpha_{0\cdot}$ for all t), then T is exponentially distributed and the R function `rexp` can be used. Table 3.1 lists R convenience functions and the associated failure time distributions. A useful overview on common parametric models for survival data is given in Klein and Moeschberger (2003, Section 2.5). A binomial experiment can be run with the function `rbinom`. We illustrate this below.

Table 3.1. R functions for simulating survival times

Distribution	R function
Exponential	`rexp`
Weibull	`rweibull`
Log-normal	`rlnorm`
Gamma	`rgamma`
Log-logistic	Use `exp` on `rlogis`

If the cause-specific hazards have been specified in such a way that a convenience function is not available, general simulation techniques will be useful. For an in-depth discussion of this topic, which is beyond the scope of this book, we refer readers to, e.g., the textbooks by Morgan (1984) and Ripley (1987); see also Rizzo (2007, Chapter 3) for an introduction to this topic using R. We have, however, chosen to briefly present the inversion method, a popular simulation technique for continuous random variables. Assume that we have specified the cause-specific hazards such that $\alpha_{0.}(t) > 0$ for all t. Then the cumulative all-cause hazard $A_{0.}(t) = \int_0^t \alpha_{0.}(u)du$ is strictly increasing (and invertible) as is the distribution function of T:

$$F(t) := P(T \leq t) = 1 - \exp(-A_{0.}(t)).$$

We write F^{-1} for the inverse of F and $A_{0.}^{-1}$ for the inverse of $A_{0.}$. Consider the transformed failure time $F(T)$. The key of the inversion method is that $F(T)$ is uniformly distributed on $[0,1]$:

$$P(F(T) \leq u) = P(T \leq F^{-1}(u)) = F(F^{-1}(u)) = u, u \in [0,1].$$

Hence, if U is a random variable with uniform distribution on $[0,1]$, then $F^{-1}(U)$ has the same distribution as T. The inversion method works as follows.

1. Compute $F^{-1}(u) = A_{0.}^{-1}(-\ln(1-u))$, $u \in [0,1]$.
2. Generate a random variable U that is uniformly distributed on $[0,1]$ (e.g., using the R function runif).
3. $F^{-1}(U)$ is the desired replicate of T.

Sometimes, the α_{0j}s are chosen in such a manner that we do not find an explicit expression for $A_{0.}^{-1}$. We may then use numerical inversion in Step 3, using the R function uniroot.

Let us now have a look at how things work out in R. The easiest and therefore very popular simulation set-up uses constant cause-specific hazards. Simulation is particularly handy, because the binomial probabilities $\alpha_{0j}/(\alpha_{01}+\alpha_{02})$ are constant over time. We generate 100 independent competing risks data with constant cause-specific hazards $\alpha_{01} = 0.3$ and $\alpha_{02} = 0.6$. We first need to generate failure times with a constant all-cause hazard $\alpha_{0.} = 0.3+0.6 = 0.9$. We have already generated such failure times in Section 2.1:

```
> event.times <- rexp(100, 0.9)
```

Now, we additionally need to generate a failure cause, either 1 or 2, for each of the 100 event times. The following code runs 100 independent binomial experiments, each of which decides on failure type 1 with probability $0.3/(0.3+0.6) = 1/3$ and, hence, with probability $1 - 1/3 = 2/3$ on failure type 2:

```
> f.cause < rbinom(100, size - 1, prob - 1/3)
```

If the binomial experiment decides on failure type 1, rbinom returns value 1 and otherwise returns 0. However, we wish our data coding to correspond to being replicates of (T, X_T) (cf. (3.3)),

```
> f.cause <- ifelse(f.cause == 0, 2, 1)
```

If the binomial experiment decided on failure type 1, `f.cause` now equals 1, whereas it otherwise equals 2.

Let us now introduce a censoring time C as in Section 2.1. We generate 100 censoring times, uniformly distributed on $[0, 5]$:

```
> cens.times <- runif(100,0,5)
```

The observable data are the censored event times $T \wedge C$,

```
> obs.times <- pmin(event.times, cens.times)
```

and the event indicator $1(T \leq C) \cdot X_T$ (cf. (3.4)),

```
> obs.cause <- c(event.times <= cens.times) * f.cause
```

In our example, we find 20 censored observations, 25 observed failures of type 1 and 55 failures of type 2,

```
> table(obs.cause)

obs.cause
 0  1  2
20 25 55
```

In the following Chapter 4, we show how we can recover the cumulative cause-specific hazards $A_{01}(t) = 0.3 \cdot t$ and $A_{02}(t) = 0.6 \cdot t$ from the data `obs.times` and `obs.cause`, and we also consider left-truncation in addition to right-censoring (cf. Section 4.2). But before this, we consider a cause-specific hazards specification, which requires some more involved coding. In particular, we look at an example, where the binomial probabilities for failure types 1 and 2 depend on the failure time T:

Assume that $\alpha_{01} = 1$ (i.e., constant), and let $\alpha_{02}(t)$ have a Weibull form,

$$\alpha_{02}(t) = \frac{a}{b^a} t^{a-1}, \ a, b > 0, \tag{3.15}$$

where a is called a shape parameter, and b is a scale parameter. Note that Weibull hazards are often written differently; e.g. Klein and Moeschberger (2003) use $\tilde{b} = b^{-a}$ as the scale parameter. Equation (3.15) is the form used in the help pages of `rweibull`, `dweibull`, `pweibull` and `qweibull`. We choose $a = 2$ and $b = 1$. The cumulative all-cause hazard is

$$A_{0\cdot}(t) = t + t^2.$$

There is no R convenience function to generate failure times with this cumulative hazard. We therefore use the inversion method. The inverse of $A_{0\cdot}$ is

$$A_{0\cdot}^{-1}(u) = \sqrt{u + 1/4} - 1/2.$$

We now generate 100 independent random variables, uniformly distributed on $[0, 1]$,

```
> my.times <- runif(100)
```

and transform them according to $A_{0.}^{-1}(-\ln(1-u))$,

```
> my.times <- -0.5 + (0.25 - log(1 - my.times))^0.5
```

Next, we decide on failure cause 1 with binomial probability $1/(1+2x)$ at a given failure time $T = x$ (cf. (3.14). An R function, which takes the simulated event times as an argument, is useful:

```
> cause1 <- function(x) {
+      out <- rbinom(length(x), 1, prob = 1 / (1 + 2 * x))
+      ifelse(out == 0, 2, 1)
+ }
```

We pass `my.times` to `cause1`, which runs the desired binomial experiments with event time-dependent probabilities,

```
> my.cause <- cause1(my.times)
```

Finally, we comment on the situation, where we do not find the inverse of A_0. analytically. Then we may employ numerical inversion, using `uniroot`. `uniroot`, applied to some function $f(x)$, searches for a root, i.e., x such that $f(x) = 0$. Recall that given a replicate of U, say $U = u$, where U is uniformly distributed on $[0, 1]$, we need to compute the transformation $A_{0.}^{-1}(-\ln(1-u)) =: t$. Then, the desired replicate of T is t. We write equivalently

$$A_{0.}(t) + \ln(1-u) = 0. \tag{3.16}$$

The idea is to use `uniroot` to find t for a given u. For illustration, we do this for the previous example with $A_{0.}(t) = t + t^2$. The following R function `generate.my.times.v2` first defines a temporary function `temp`, which is the cumulative all-cause hazard plus some y. Because of (3.16), we use `y = log(1 - u)` below. `temp` is called by `uniroot`. `uniroot` searches for a root on a prespecified interval. We search on the interval from 0 to `max.int`, which has to be passed to `generate.my.times.v2` together with the number `n` of desired replicates. `uniroot` also requires the function values (of the function `temp`) at the endpoints to be of opposite signs (or zero), which is tested before calling `uniroot`:

```
> generate.my.times.v2 <- function(n, max.int) {
+      temp <- function(x,y) { return(x + x^2 + y) }
+      stime <- NULL
+      i <- 1
+
+      while(length(stime) < n) {
+          u <- runif(1)
+          ## If endpoints are of opposite sign, call uniroot:
+          if (temp(0, log(1 - u)) *
+              temp(max.int, log(1 - u)) < 0) {
```

```
+              res <- uniroot(temp, c(0, max.int),
+                                tol = 0.0001, y = log(1 - u))
+              stime[i] <- res$root
+              i <- i + 1
+           }
+           else cat("Values at endpoints not
+                    of opposite signs. \n")
+         }
+         return(stime)
+ }
```

Running `my.times <- generate.my.times.v2(n=100, max.int=99)` gives us the desired 100 independent replicates of T with cumulative all-cause hazard $A_{0.}(t) = t + t^2$; here, we have asked `uniroot` to search on the interval $[0, 99]$.

We finally note that competing risks simulations in the literature are often based on an empirically unverifiable data structure (Beyersmann et al., 2009), namely the latent failure time model. This model and its inherent difficulties are briefly discussed in Section 3.3.

3.3 The latent failure time model

Some readers may be familiar with competing risks as arising from risk-specific latent times. This is the latent failure time approach to competing risks. As explained below, the latent failure time model of competing risks is 'larger' than the multistate model of Section 3.1. This also explains the difficulties that come with assuming latent times. The interpretation of the latent times will often be awkward, any independence or dependence between them can usually not be verified empirically. Nevertheless, the concept of underlying risk-specific times has remained a vivid notion in the context of competing risks. In fact, sometimes in the literature, competing risks are considered to arise from latent times, but as the observable data are as in Sections 3.1 and 3.2, subsequent analyses do not rely on the assumption of latent failure times.

We have therefore chosen to present briefly the latent failure time model, too. To be specific, assume that there are random variables $T^{(1)}, T^{(2)} \in [0, \infty)$. The connection to the multistate-type data (3.2) and (3.3) is

$$T = T^{(1)} \wedge T^{(2)} \text{ and } X_T = 1 \iff T^{(1)} < T^{(2)}.$$

It is important to note that the data (T, X_T) arising from the multistate model of Figure 3.1 are observable (save for left-truncation and right-censoring). $T^{(1)}, T^{(2)}$, however, are not observable. Only their minimum and the indicator $\mathbf{1}(T^{(1)} < T^{(2)})$ are observable.

The latent failure time model has been criticized mainly for three reasons. First, whether $T^{(1)}$ and $T^{(2)}$ are independent or dependent cannot be verified from the observable data, even in the absence of right-censoring and left-truncation. We refer to Kalbfleisch and Prentice (2002, Section 8.2.4) for a proof and an in-depth discussion; see also Tsiatis (1975). The basic idea is that one can find joint distributions of $T^{(1)}$ and $T^{(2)}$, where $T^{(1)}$ and $T^{(2)}$ are dependent, that give rise to the same likelihood which also arises assuming independence. This hampers any statistical analysis of the latent times, which will usually need to make an assumption about any dependence between $T^{(1)}$ and $T^{(2)}$. Typically, for tractability, $T^{(1)}$ and $T^{(2)}$ are assumed to be independent. This assumption is, just as typically, considered to be rather strong and restrictive.

The non-identifiability of the joint distribution is a serious statistical problem, *if* one assumes the existence of latent times. But even more important are two conceptual concerns:

Because the latent times are unobservable, there is something hypothetical about them. As a consequence, latent times have been criticized for lack of plausibility (e.g., Prentice et al., 1978). We illustrate this point for the situation of hospital stay in a real data analysis in Section 4.3.1. We remark, however, that there are special situations where the latent times can be given a physical interpretation: a technical device may be made of a number of components such that its lifetime is determined by the smallest lifetime of the components. A very simple example is a chain of light bulbs that are connected in series. The chain fails when the first light bulb fails. It is important to note, though, that analysing the lifetime of a chain of light bulbs would not need to require the existence of latent times. The lifetime may very well be analysed using the hazard-based techniques of this book, making no assumption on latent times whatsoever.

Third, and perhaps most important, it is disputable whether the latent failure time point of view constitutes a fruitful approach to answer questions of the original subject matter. Aalen (1987) criticizes the issue of non-identifiability of the joint distribution of $T^{(1)}$ and $T^{(2)}$ as being an 'artificial problem'. We illustrate in a real data analysis in Section 4.3.1 that assuming latent times does not provide further insight as compared to a multistate-type analysis based on cause-specific hazards. But the analysis based on $T^{(1)}$ and $T^{(2)}$ has to cope with an awkward interpretation of the latent times, has to make an unverifiable assumption on their statistical dependence, *and* its interpretation turns out to be less straightforward.

The remainder of the competing risks part of this book illustrates that competing risks data may very well be analysed without assuming risk specific latent times and, hence, without needing to cope with the difficulties mentioned above. This is also of relevance for analysing multistate data which are realized as a nested sequence of competing risks experiments: we do not need to assume that individuals are exposed to a nested sequence of latent, hypothetical times in order to analyse such data.

The attraction of the latent failure time approach despite these difficulties probably lies in the fact that it suggests a way to answer 'what if' questions: if we assume that $T^{(1)}$ is the time until event 1 in a world where events of the competing type 2 have been prevented, and if we are able to estimate $P(T^{(1)} > t)$ (e.g., under independence of $T^{(1)}$ and $T^{(2)}$), then the latent failure time approach provides a way to predict the survival probability under 'cause removal'. Gail (1982) discusses such calculations as a major application of the latent failure time model.

Both competing risks analyses and considerations on the effect of cause removal can be traced back to Daniel Bernoulli's *Essai d'une nouvelle analyse de la mortalité causée par la petite vérole, et des avantages de l'inoculation pour la prévenir*, read before the Royal Academy of Sciences in Paris in 1760 and finally published in 1766. A reprint of the paper is in Bernoulli (1982) and an English translation (reprinted in parts in Bernoulli and Blower (2004)) is in Bradley (1971). The aim of Bernoulli's analysis was to study the impact of vaccination against smallpox on survival. In Bernoulli's time, smallpox was endemic and a major cause of death; vaccination against smallpox and subsequent adverse events were the topic of a controversial public health debate (e.g., Bernoulli and Blower, 2004).

Bernoulli attacked the problem by defining (time-constant) hazards of infection, of dying due to smallpox or due to other reasons and of becoming immune, i.e., becoming infected and surviving smallpox. He then studied the impact of mass vaccination by substituting zero for the infection hazard.

It is important to note that Bernoulli's approach addressed the question of 'What if smallpox were eradicated?' *without* referring to latent failure times. The advantage of Bernoulli's approach is obvious. The reasoning neither builds on hypothetical times with an awkward interpretation nor does it require us to make an independence assumption between these times. What it does assume, however, is that the remaining hazards are the same in a world with or without smallpox.

An excellent modern discussion of Bernoulli's paper is given by Dietz and Heesterbeek (2002). They also discuss that Bernoulli's basic model was actually an illness-death model (cf. Section 2.2.4), the intermediate 'illness'-state corresponding to 'being immune'. Constructing *partial* processes like Bernoulli by substituting zero for some hazards is, e.g., discussed by Andersen et al. (1993, Section IV.4.1.6). Andersen et al. (2002) discuss investigating different hypothetical scenarios of modified hazards in a kind of sensitivity analysis. Although speculative, these models do not require a latent failure time model.

In summary, it is our impression that latent failure times are usually not needed in the analysis of competing risks data. In contrast, assuming latent times rather appears to cause confusion and to create artificial problems than to contribute to an understanding of the subject matter. Although logically feasible, the concept of latent times is not convincing, except for special cases such as a technical device whose single components have a physical and functional interpretation. Except for such special cases, the latent failure time model is philosophically rather 'expensive', assuming, e.g., that a human being is equipped with a large enough reservoir of latent times for any competing risks situation that the individual might face.

3.4 Exercises

1. Show that the definition of the cause-specific hazards implies that

$$P(T \leq t, X_T = 1) = \int_0^t P(T \geq u)\alpha_{01}(u)\,du.$$

Show that

$$P(T \leq t) = 1 - \exp(-A_0.(t)) = \int_0^t P(T \geq u)\alpha_0.(u-)\,du.$$

2. *The competing risks simulation algorithm works*: Simulate competing risks data with cause-specific hazard of interest following a Weibull distribution with shape parameter a equal to 3 and scale parameter $b = 2$,

$$\alpha_{01}(t) = \frac{3}{8}t^2,$$

and competing cause-specific hazard defined as

$$\alpha_{02}(t) = \frac{1.8}{t+2}.$$

For completely observed data, the usual event proportions

$$\frac{\text{\# of type } j \text{ events in } [0,t]}{\text{\# of individuals}}$$

estimate $P(T \leq t, X_T = j)$. Compare the event proportions with the true quantities.

3. *Latent failure time 'counter example'*: Recall the latent failure time model of Section 3.3 with joint survival function $P(T^{(1)} > t_1, T^{(2)} > t_2) = Q(t_1, t_2)$. Consider

   (a) $Q(t_1, t_2) = \exp\{1 - \gamma_1 t_1 - \gamma_2 t_2 - \exp[\gamma_{12}(\gamma_1 t_1 + \gamma_2 t_2)]\}$,

   (b) $Q(t_1, t_2) = \exp\{1 - \gamma_1 t_1 - \gamma_2 t_2 - \dfrac{\gamma_1 e^{\gamma_{12}(\gamma_1+\gamma_2)t_1} + \gamma_2 e^{\gamma_{12}(\gamma_1+\gamma_2)t_2}}{\gamma_1 + \gamma_2}\}$,

   where $\gamma_1, \gamma_2 > 0$ and $\gamma_{12} > -1$. For both (a) and (b) compute (and compare) the cause-specific hazards and the hazards of the marginal survival functions $P(T^{(1)} > t)$ and $P(T^{(2)} > t_2)$. Which of the models (a) and (b) is an independent latent failure time model? (Hint: See Section 8.2.4 of Kalbfleisch and Prentice (2002).)

4

Nonparametric estimation

We introduce the key nonparametric estimators, the Nelson-Aalen estimator of the cumulative cause-specific hazards and the Aalen-Johansen estimator of the cumulative incidence functions, in Section 4.1. We analyse the simulated data of Section 3.2 in Section 4.2; this section also introduces in detail the functionality offered by the R packages mvna, etm, cmprsk, and survival for nonparametric estimation in a competing risks model. We analyse a real data example in Section 4.3; this section emphasizes interpretation of competing risks analyses. The usual nonparametric estimators for standard single endpoint survival analysis, i.e., the Nelson-Aalen estimator of the cumulative all-cause hazard and the Kaplan-Meier estimator of the survival function, are included in the present account.

We assume that there are n individuals under study with competing risks data arising from n independent replicates of a competing risks process as in Section 3.1, subject to independent right-censoring and left-truncation as in Section 2.2.2.

4.1 The Nelson-Aalen estimator and the Aalen-Johansen estimator

The cause-specific hazards $\alpha_{0j}(t)$, $j = 1, 2$ are the key quantities of the competing risks model introduced in Section 3.1, and they completely determine the stochastic behaviour of the competing risks process: specification of the α_{0j}s in Section 3.2 sufficed to generate competing risks data (except, of course, for an outward censoring or truncation mechanism.) In this section, we show that the Nelson-Aalen estimators of the *cumulative* cause-specific hazards $A_{0j}(t) = \int_0^t \alpha_{0j}(u)\mathrm{d}u$, $j = 1, 2$, are the key building block for non-parametric estimation in competing risks.

Note that we estimate the cumulative quantities rather than the α_{0j}s themselves, as the latter is a much harder problem. Like a density function, the

α_{0j}s can be virtually any nonnegative function. However, such as the cumulative distribution function, the cumulative hazard function can be estimated straightforwardly. And, as we explained in Chapter 2, estimating the A_{0j}s also 'suffices' in the following sense: the standard estimators of the survival function and of the cumulative incidence function are deterministic functions — product integrals — of the Nelson-Aalen estimators. Although beyond the technical level of this book, we also note that this fact can also be used to approximate the distribution of the probability estimators via the functional delta method, given pertinent results of the computationally simpler Nelson-Aalen estimators; see, e.g., Aalen et al. (2008). This further underlines the key role played by the Nelson-Aalen estimators.

We consider n individuals under study with individual competing risks process $(X_t^{(i)})_{t \geq 0}$, $X_t^{(i)} \in \{0, 1, 2\}$, $i = 1, 2, \ldots, n$. The individual failure time is T_i with failure cause $X_{T_i}^{(i)} =: X_{T_i}$. Observation of the individual competing risks data is subject to a right-censoring time C_i and possibly also to a left-truncation time L_i. We assume that right-censoring and left-truncation are independent as explained in Section 2.2.2. We introduce some notation connected to occupation of the states and possible transitions between them of the competing risks multistate picture in Figure 3.1.

The ith at-risk process is

$$Y_{0;i}(t) := \mathbf{1}(L_i < t \leq T_i \wedge C_i) \tag{4.1}$$

In words and in the absence of left-truncation, $Y_{0;i}(t)$ equals 1, as long as individual i is in the initial state 0 of Figure 3.1 *and* under observation just prior to time t. If $Y_{0;i}(t) = 1$, individual i may be *observed* to experience one of the competing events at time t, or individual i may be censored at t or remain under observation in state 0. In the presence of a left-truncation time L_i, which we assume to be less than C_i, observation of individual i starts with time L_i, and the individual is considered to be at risk of experiencing an *observable* competing event or of being censored only after L_i. Note that individual i, $i = 1, \ldots, n$ is an individual *under study*. That is, the individual's failure time T_i is greater than the potential left-truncation time L_i. In the presence of left-truncation, there may be individuals who never enter the study, that is, their failure time is less than their left-truncation time, but these individuals are not part of the n individuals under study.

Individual i may experience one of the two competing events 1 and 2 at time T_i, modelled by making the transition from state 0 to state 1 (if competing event 1 occurs) or to state 2 (if competing event 2 occurs). We count which event types we observe over the course of time:

$$N_{01;i}(t) := \mathbf{1}(T_i \wedge C_i \leq t, L_i < T_i \leq C_i, X_{T_i} = 1) \tag{4.2}$$

equals 1 if we observe individual i to make the transition from state 0 to state 1 during the time interval $[0, t]$. Otherwise, $N_{01;i}(t) = 0$. Analogously,

$$N_{02;i}(t) := \mathbf{1}(T_i \wedge C_i \leq t, L_i < T_i \leq C_i, X_{T_i} = 2) \tag{4.3}$$

equals 1 if we observe a $0 \to 2$ transition for individual i during the time interval $[0, t]$, and $N_{02;i}(t) = 0$ otherwise.

We aggregate over all individuals $i = 1, \ldots, n$. The number of individuals to be observed at risk in the initial state 0 just prior to time t is

$$Y_0(t) := \sum_{i=1}^{n} Y_{0;i}(t), \tag{4.4}$$

and the number of individuals observed to make the $0 \to j$ transition during the time interval $[0, t]$ is

$$N_{0j}(t) := \sum_{i=1}^{n} N_{0j;i}(t). \tag{4.5}$$

We also write

$$N_{0\cdot}(t) := N_{01}(t) + N_{02}(t) \tag{4.6}$$

for the number of observed transitions out of the initial state 0 during the time interval $[0, t]$, and we write

$$\Delta N_{0j}(t) := N_{0j}(t) - N_{0j}(t-) \text{ and } \Delta N_{0\cdot}(t) := N_{0\cdot}(t) - N_{0\cdot}(t-) \tag{4.7}$$

for the increments of $N_{0j}(t)$, i.e., the number of $0 \to j$ transitions observed *exactly* at time t, and for the increments of $N_{0\cdot}(t)$, respectively. The connection between these counting processes and the competing risks multistate picture of Figure 3.1 is illustrated in Figure 4.1.

Figure 4.1 also illustrates the *increments* of the Nelson-Aalen estimator $\widehat{A}_{0j}(t)$ of $A_{0j}(t) = \int_0^t \alpha_{0j}(u)\mathrm{d}u$, $j = 1, 2$. To motivate the estimator, recall from Equation (3.5) that $\alpha_{01}(t)\,\mathrm{d}t$, say, is an infinitesimal conditional transition probability

$$\alpha_{01}(t)\mathrm{d}t = P(T \in \mathrm{d}t, X_T = 1 \,|\, T \geq t).$$

If we observe no $0 \to 1$ transition at t (i.e., $\Delta N_{01}(t) = 0$), we estimate the increment $\alpha_{01}(t)\mathrm{d}t$ of the cumulative $0 \to 1$ hazard as 0. If we do observe $0 \to 1$ transitions at t (i.e., $\Delta N_{01}(t) > 0$), we estimate this conditional transition probability as the ratio of the number $\Delta N_{01}(t)$ of $0 \to 1$ transitions divided by the number $Y_0(t)$ at risk just prior to the transition time t. We proceed in an analogous manner for the competing $0 \to 2$ transition. Summing up over these increments yields the Nelson-Aalen estimators:

$$\widehat{A}_{0j}(t) := \sum_{T_i \wedge C_i \leq t} \frac{\Delta N_{0j}(T_i \wedge C_i)}{Y_0(T_i \wedge C_i)}, \ j = 1, 2. \tag{4.8}$$

An estimator of the variance of $\widehat{A}_{0j}$ can be derived using martingale theory. (See Section 2.2.1 on the connection between counting the number of observed events and what a martingale is.)

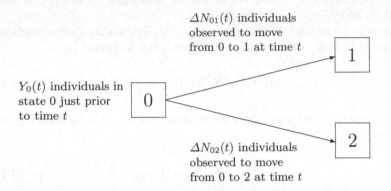

Fig. 4.1. Competing risks multistate model with number at risk just prior to t and number of observed event-specific transitions at t.

The estimator is

$$\widehat{\sigma}_{0j}^2(t) := \sum_{T_i \wedge C_i \leq t} \frac{\Delta N_{0j}(T_i \wedge C_i)}{Y_0^2(T_i \wedge C_i)}, \ j = 1, 2. \tag{4.9}$$

The Nelson-Aalen estimators are asymptotically normally distributed (e.g., Andersen et al., 1993, Section IV.1.2). We also note that they are asymptotically independent. The limit distribution can be used to construct approximate $100 \cdot (1-\alpha)\%$ confidence intervals of $\widehat{A}_{0j}(t)$ at a given time point t, $\alpha \in (0,1)$. The small sample properties of such an interval may be improved by using a log-transformation (Bie et al., 1987), resulting in the confidence interval

$$\widehat{A}_{0j}(t) \exp \left(\pm z_{1-\alpha/2} \widehat{\sigma}_{0j}(t) / \widehat{A}_{0j}(t) \right), \tag{4.10}$$

where $z_{1-\alpha/2}$ is the $1 - \alpha/2$ quantile of the standard normal distribution; it can be obtained in R applying `qnorm` to the value of $1 - \alpha/2$. Note that the confidence interval (4.10) is pointwise in nature, i.e., for a given time point t; it does not yield a $100 \cdot (1-\alpha)\%$ confidence band for $\widehat{A}_{0j}$ as a function of time. Confidence bands are, e.g., discussed in Andersen et al. (1993, Section IV.1.3); a formula for pointwise confidence intervals after transformation is given in Andersen et al. (1993, p. 208).

The event-specific Nelson-Aalen estimators sum up to the Nelson-Aalen estimator of the cumulative all-cause hazard $A_{0\cdot}(t)$:

$$\widehat{A}_{0\cdot}(t) := \widehat{A}_{01}(t) + \widehat{A}_{02}(t) = \sum_{T_i \wedge C_i \leq t} \frac{\Delta N_{0\cdot}(T_i \wedge C_i)}{Y_0(T_i \wedge C_i)}, \tag{4.11}$$

which has already been introduced in Chapter 2 (cf. Equation (2.14)). An estimator of the variance of $\widehat{A}_0.(t)$ is

$$\widehat{\sigma}_0^2.(t) := \sum_{T_i \wedge C_i \leq t} \frac{\Delta N_0.(T_i \wedge C_i)}{Y_0^2(T_i \wedge C_i)}, \tag{4.12}$$

paralleling Equation (4.9). A pointwise confidence interval can be found analogously to (4.10).

In Chapter 2, we also found that the Kaplan-Meier estimator is a deterministic function of the all-cause Nelson-Aalen estimator $\widehat{A}_0.(t)$ (cf. Equation (2.15)). Because of Equation (4.11), the Kaplan-Meier estimator is also a deterministic function of both cause-specific Nelson-Aalen estimators. In competing risks, the Kaplan-Meier estimator estimates the survival function of the waiting time T in the initial state 0 of Figures 3.1 and 4.1:

$$\widehat{P}(T > t) := \prod_{T_i \wedge C_i \leq t} \left(1 - \frac{\Delta N_0.(T_i \wedge C_i)}{Y_0(T_i \wedge C_i)} \right). \tag{4.13}$$

We also write

$$\widehat{P}(T > t-) := \prod_{T_i \wedge C_i < t} \left(1 - \frac{\Delta N_0.(T_i \wedge C_i)}{Y_0(T_i \wedge C_i)} \right) \tag{4.14}$$

for the value of the Kaplan-Meier estimator just prior to t.

An estimator of the variance of the Kaplan-Meier estimator is:

$$\widehat{VAR}\left(\widehat{P}(T > t)\right) := \left(\widehat{P}(T > t)\right)^2 \cdot$$
$$\sum_{T_i \wedge C_i \leq t} \frac{\Delta N_0.(T_i \wedge C_i)}{Y_0(T_i \wedge C_i) \cdot (Y_0(T_i \wedge C_i) - \Delta N_0.(T_i \wedge C_i))} \tag{4.15}$$

Equation (4.15) is the famous Greenwood formula. We note that an alternative variance estimator can be derived by substituting $\widehat{\sigma}_0^2.(t)$ from (4.12) for the last term on the right hand side of (4.15). Conversely, the last term on the right hand side of (4.15) can be seen as an alternative variance estimator of the cumulative all-cause hazard $\widehat{A}_0.(t)$. We use the estimators (4.12) and (4.15) in compliance with standard practice (e.g., Andersen et al. (1993) and Klein and Moeschberger (2003)).

Asymptotic normality of the Kaplan-Meier estimator (e.g., Andersen et al., 1993, Section IV.3.2), can be used to construct approximate $100 \cdot (1 - \alpha)\%$ confidence intervals. Again, it is advisable to use transformations to improve the small sample properties. We consider the log-minus-log transformation, resulting in the confidence interval

$$\widehat{P}(T > t)^{\exp\left\{\pm z_{1-\alpha/2}\sqrt{\widehat{VAR}(\widehat{P}(T>t))}\,/\,(\widehat{P}(T>t)\ln\widehat{P}(T>t))\right\}}. \tag{4.16}$$

Note that the log-minus-log transformation for the Kaplan-Meier estimator corresponds to the log-transformation for the Nelson-Aalen estimator, because $\ln(P(T > t)) = \ln \exp(-A(t)) = -A(t)$. The transformation also ensures that the confidence interval (4.16) is always contained in $[0, 1]$. Like the confidence intervals (4.10) for the Nelson-Aalen estimates, the interval (4.16) should be given a pointwise interpretation.

We finally derive an estimator of the cumulative incidence functions from the Kaplan-Meier estimator of the survival function. Recall from (3.12) that the cumulative incidence functions $P(T \leq t, X_T = 1)$ and $P(T \leq t, X_T = 2)$ add up to the all-cause distribution function $P(T \leq t)$. An estimator of $P(T \leq t)$ is 1 minus the Kaplan-Meier estimator of the survival function. An easy algebraic calculation shows that

$$1 - \widehat{P}(T > t) = 1 - \prod_{T_i \wedge C_i \leq t} \left(1 - \frac{\Delta N_{0\cdot}(T_i \wedge C_i)}{Y_0(T_i \wedge C_i)} \right)$$

$$= \sum_{T_i \wedge C_i \leq t} \widehat{P}(T > (T_i \wedge C_i) -) \cdot \frac{\Delta N_{0\cdot}(T_i \wedge C_i)}{Y_0(T_i \wedge C_i)}. \quad (4.17)$$

(One easily verifies (4.17) by checking the increments of the respective representations.) The right hand side of (4.17) splits into the cumulative incidence function estimators, recalling that $N_{0\cdot} = N_{01} + N_{02}$:

$$\widehat{P}(T \leq t, X_T = j) := \sum_{T_i \wedge C_i \leq t} \widehat{P}(T > (T_i \wedge C_i) -) \cdot \frac{\Delta N_{0j}(T_i \wedge C_i)}{Y_0(T_i \wedge C_i)}, \quad j = 1, 2$$

$$(4.18)$$

Estimators (4.18) are special cases of the Aalen-Johansen estimator for general time-inhomogeneous Markov processes (cf. Section 2.2.4).

It is worthwhile to consider for a moment the different *interpretations* of the Kaplan-Meier estimator of (4.13) and of one minus the Kaplan-Meier estimator in (4.17); the latter will directly lead to the correct interpretation of the cumulative incidence function estimator: Equation (4.13) is the standard representation of the Kaplan-Meier estimator. It is a product over empirical conditional survival probabilities: $\Delta N_{0\cdot}(T_i \wedge C_i)/Y_0(T_i \wedge C_i)$ is the empirical probability to fail at time $T_i \wedge C_i$, given that one has not failed before. One minus this quantity is the empirical probability of surviving $T_i \wedge C_i$ conditional on not having failed before. The product over these terms for all times $T_i \wedge C_i \leq t$ results in the empirical probability of surviving (i.e., not failing) up to and including time t.

One minus this empirical survival probability is the empirical probability of failing up to and including time t. Now consider (4.17). The right hand side is the sum over all empirical probabilities of failing at time $T_i \wedge C_i$, where we sum up over all $T_i \wedge C_i \leq t$: $\widehat{P}(T > (T_i \wedge C_i) -)$ is the probability of not having failed prior to $T_i \wedge C_i$; this probability is multiplied with the conditional probability of failing at time $T_i \wedge C_i$, given that one has not failed before.

This reasoning should lead to a cumulative incidence function estimator for event type 1, say, if we do not sum up over all empirical probabilities of failing at time $T_i \wedge C_i$, but over all probabilities of failing *from cause* 1 at $T_i \wedge C_i$: we multiply the probability of not having failed prior to $T_i \wedge C_i$ with conditional probability $\Delta N_{01}(T_i \wedge C_i)/Y_0(T_i \wedge C_i)$ of failing *from cause* 1 at $T_i \wedge C_i$, given that one has not failed before. This is the desired estimator (4.18). Also note that (4.18) is the empirical analogue of the right hand side of (3.11).

We finally consider estimating the variance of the cumulative incidence function estimators. We define

$$\widehat{P}(T \leq t, X_T = j \mid T > s) := \sum_{s < T_i \wedge C_i \leq t} \frac{\widehat{P}(T > (T_i \wedge C_i) -)}{\widehat{P}(T > s)} \cdot \frac{\Delta N_{0j}(T_i \wedge C_i)}{Y_0(T_i \wedge C_i)}$$

(4.19)

for $j = 1, 2$ and $s < t$. $\widehat{P}(T \leq t, X_T = j \mid T > s)$ estimates the probability of failing from cause j up to and including time t, given that one has survived time s. Note that there is no truly conceptual difference between (4.18) and (4.19): the latter considers the competing risks process starting at time $s > 0$ and, hence, only considers those still at risk in state 0 at that time, whereas the former starts at time 0. A Greenwood-type estimator of the variance is:

$$\widehat{\mathrm{VAR}}(\widehat{P}(T \leq t, X_T = j)) =$$

$$\sum_{s \leq t} \frac{\left(\widehat{P}(T \leq t, X_T = j) - \widehat{P}(T \leq s, X_T = j)\right)^2}{Y_0(s) - \Delta N_{0\cdot}(s)} \Delta \widehat{A}_{0\cdot}(s) +$$

$$\frac{\widehat{P}(T > s-)^2}{Y_0(s)^3} \Delta N_{0j}(s) \cdot \left\{ Y_0(s) - \Delta N_{0j}(s) \right.$$

$$\left. - 2(Y_0(s) - \Delta N_{0\cdot}(s)) \cdot \frac{\widehat{P}(T \leq t, X_T = j) - \widehat{P}(T \leq s, X_T = j)}{\widehat{P}(T > s)} \right\}$$

(4.20)

This estimator is formally derived in Andersen et al. (1993, Section IV.4). In the absence of a competing event state 2, formula (4.20) coincides with (4.15). Performance of variance estimators for the estimated cumulative incidence function has been investigated with about a 15 year delay as compared to the standard survival case; see, e.g., Braun and Yuan (2007). The preferred variance estimator is (4.20); see Allignol et al. (2010).

Following a suggestion by Lin (1997), we use the transformation $x \mapsto \ln(-\ln(1-x))$ to construct approximate $100 \cdot (1-\alpha)\%$ confidence intervals for $\widehat{P}(T \leq t, X_T = j)$. These can again be justified by an asymptotic normality result (e.g., Andersen et al., 1993, Section IV.4.2). The transformation corresponds to the log-minus-log transformation of (4.16), applied to $1 - \widehat{P}(T \leq t, X_T = j)$, and resulting in the confidence intervals

$$1 - (1 - \widehat{P}(T \leq t, X_T = j))^{\exp\left(\mp z_{1-\alpha/2} \frac{\sqrt{\widehat{\mathrm{VAR}}(\widehat{P}(T \leq t, X_T = j))}}{(1 - \widehat{P}(T \leq t, X_T = j)) \cdot \ln(1 - \widehat{P}(T \leq t, X_T = j))}\right)}, \quad (4.21)$$

for $j = 1, 2$. As before, the confidence interval should be given a pointwise interpretation. The confidence intervals (4.21) are always contained in $[0, 1]$. In fact, $P(T \leq t, X_T = j)$ always lies in $[0, P(X_T = j)]$, where $P(X_T = j) = \lim_{t \to \infty} P(T \leq t, X_T = j)$ is the expected proportion of individuals who fail from cause j. However, we usually do not know $P(X_T = j)$, and this quantity will not even be nonparametrically estimable with most right-censored data (cf. Section 2.3).

We now explain how to perform the analyses in R. We first consider how to recover the cumulative cause-specific hazards, from which we generated data in Section 3.2. Next, we consider two real data examples. The analysis of the simulated competing risks data is focused on presenting the tools provided by R and also on how well we may recover the theoretical quantities. The first real data example focuses on interpreting a competing risks analysis, taking advantage of the R tools provided earlier. This highlights and explains some of the subtleties that are common in competing risks. The second data example briefly illustrates how left-truncation is easily incorporated in the present framework.

4.2 Analysis of simulated competing risks data

Cumulative cause-specific hazards

Recall from Section 3.2 that we generated data obs.times and obs.cause of 100 individuals from cumulative cause-specific hazards $A_{01}(t) = 0.3 \cdot t$ and $A_{02}(t) = 0.6 \cdot t$ and an independent censoring time that was uniformly distributed on $[0, 5]$. Later, we additionally consider left-truncation, too.

To begin, we estimate the cumulative hazards, using the mvna package (Allignol et al., 2008). First, we need to describe the competing risks multistate model of Figures 3.1 and 4.1. We define a matrix of logical values indicating the possible transition types within our multistate model:

```
> tra <- matrix(FALSE, ncol = 3, nrow = 3)
> dimnames(tra) <- list(c("0", "1", "2"), c("0", "1", "2"))
> tra[1, 2:3] <- TRUE
> tra

      0     1     2
0 FALSE  TRUE  TRUE
1 FALSE FALSE FALSE
2 FALSE FALSE FALSE
```

tra tells us that an individual may move from state 0 to state 1 and from state 0 to state 2. Backward transitions are not possible. Also, the values on the diagonal are FALSE: 'transitions' from one state into itself are not modelled. There is no need to model such 'transitions'. Individuals who do not make a transition into one of the two competing even states at time t, say, remain in the initial state 0 at t.

Next, we aggregate obs.times and obs.cause into a data frame my.data. For all individuals, my.data has one row for each observed transition and one row for a censoring event. In a competing risks model, an individual is either observed to experience a competing event or is censored. Thus, my.data has as many rows as there are individuals. We also define an individual ID variable and mark censoring events by a value cens:

```
> id <- seq_along(obs.cause)
> from <- rep(0, length(obs.cause))
> to <- as.factor(ifelse(obs.cause == 0, "cens", obs.cause))
> my.data <- data.frame(id, from, to, time = obs.times)
```

We have a quick look at the new data frame:

```
> head(my.data)
```

```
  id from   to      time
1  1    0    2 0.1450839
2  2    0    2 2.3988344
3  3    0    1 0.4447219
4  4    0 cens 0.3656826
5  5    0    1 0.6050701
6  6    0    2 0.8025029
```

Individual 3 experiences competing event 1 at time 0.4447219, individual 1 experiences competing event 2 at time 0.1450839, and individual 4 is censored in the initial state 0 at time 0.3656826

We are now set to estimate the cumulative cause-specific hazards using the function mvna of the mvna package. The function mvna requires as arguments a data frame as created above, the state names which will be the same as in Figure 3.1, a matrix defining the possible transitions, and the name marking censored observations:

```
> library(mvna)
> my.nelaal <- mvna(my.data, c("0", "1", "2"), tra, "cens")
```

The value returned by mvna is a list with components named after the possible transitions, in our example "0 1" and "0 2". These components are data frames: my.nelaal[["0 1"]]$na are the estimates $\widehat{A}_{01}(t)$ at times t as in my.nelaal[["0 1"]]$time; the values of the variance estimator (4.9) are in my.nelaal[["0 1"]]$var.aalen.

```
> head(my.nelaal[["0 1"]][, c( "time","na","var.aalen")])
```

```
                           time        na      var.aalen
0                    0.00000000 0.00000000 0.0000000000
0.0136922217077679   0.01369222 0.01000000 0.0001000000
0.0170758542501264   0.01707585 0.01000000 0.0001000000
0.0241606461349875   0.02416065 0.01000000 0.0001000000
0.0353729078132245   0.03537291 0.02030928 0.0002062812
0.0369639217387885   0.03696392 0.02030928 0.0002062812
```

We note that the variance estimates may be biased when the risk set is small, say, less than five individuals (Klein, 1991). With competing risks data that are only subject to right-censoring but not to left-truncation, this usually only happens for late event times.

We plot the Nelson-Aalen estimates for both possible transitions along with the log-transformed confidence intervals of (4.10):

```
> xyplot(my.nelaal, strip=strip.custom(bg="white"),
+        ylab="Cumulative Hazard", lwd=2)
```

The output is displayed in Figure 4.2 where we have also included the theoretical quantities $A_{01}(t) = 0.3t$ and $A_{02}(t) = 0.6t$ for comparison. In this

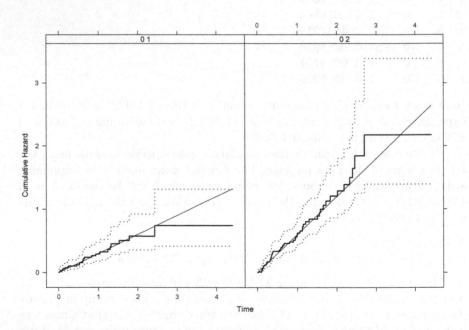

Fig. 4.2. *Simulated data.* The step functions are the Nelson-Aalen estimates $\widehat{A}_{01}(t)$ (left) and $\widehat{A}_{02}(t)$ (right) with log-transformed confidence intervals for the data **obs.times** and **obs.cause** that were generated in Section 3.2. The linear functions are the true cumulative hazards $A_{01}(t)$ and $A_{02}(t)$.

example, the Nelson-Aalen estimates approximate the true cumulative hazards quite well beyond the 75th quantile,

```
> qexp(0.75, rate = 0.9)
```

```
[1] 1.540327
```

Survival function

We now consider estimating the survival function $P(T > t)$ of the waiting time T in the initial state 0. The Kaplan-Meier estimator (4.13) may be computed from the value returned by mvna, as it is a deterministic function of the all-cause Nelson-Aalen estimator $\widehat{A}_{0.}(t) = \widehat{A}_{01}(t) + \widehat{A}_{02}(t)$ (cf. (4.11)). A more convenient way is provided by the survfit-function of the survival package:

```
> my.fit.surv <- survfit(Surv(time, to != "cens") ~ 1,
+                         data = my.data, conf.type = "log-log")
```

The first argument to survfit, i.e., Surv(time, to != "cens") creates a 'survival object' from event times my.data$time; see also Section 2.1. An event time is considered as having been observed whenever my.data$to does *not* equal "cens". The last argument requires confidence intervals to be computed from the log-minus-log transformation (cf. (4.16)). survfit returns a 'survfit object'. my.fit.surv$surv are the values of the Kaplan-Meier estimator at time points my.fit.surv$time.

The result can be plotted:

```
> plot(my.fit.surv, xlab = "Time", ylab = "Survival
+         Probability", mark.time= FALSE, lwd = 2)
```

mark.time = FALSE requests that the Kaplan-Meier curve is *not* marked at censoring times, which are also not an event time. We have chosen not to mark these times in compliance with Figure 4.2. The Kaplan-Meier curve is shown in Figure 4.3 together with its theoretical counterpart, added via curve(exp(-0.9 * x), add = TRUE). Also note that plot applied to my.fit.surv plots pointwise confidence intervals as given in (4.16), based on the Greenwood variance estimator (4.15). However, an estimator of the variance of the Kaplan-Meier estimator is not explicitly returned by the survfit-function. This functionality is offered by the etm package described below.

Estimating the survival function from a multistate perspective

Using the survival package is the standard way in R to compute the Kaplan-Meier estimator. However, the package does not provide for estimating probabilities in more complex multistate models. This functionality is offered by the etm package (Allignol et al., 2011a). Because the standard survival situation can be displayed as a very simple multistate model (see Figure 2.1) we have

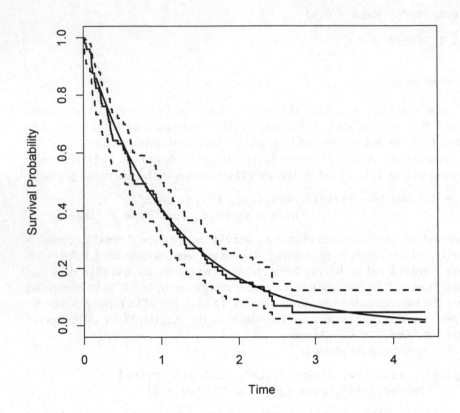

Fig. 4.3. *Simulated data.* Step functions: Kaplan-Meier estimate $\widehat{P}(T > t)$ with log-minus-log transformed confidence intervals (4.16) for the data `obs.times` and `obs.cause` that were generated in Section 3.2. Smooth line: theoretical survival function.

chosen to explain computation of the Kaplan-Meier estimator using `etm`, too. As described before, one way to do this is by combining the competing event states 1 and 2 of Figure 3.1 into one single absorbing state as in Figure 2.1, which we also call state 1. First, we again define a matrix of logical values with now only one possible transition type:

```
> tra.km <- matrix(FALSE, ncol=2, nrow=2)
> dimnames(tra.km) <- list(c("0", "1"), c("0", "1"))
> tra.km[1, 2] <- TRUE
> tra.km
```

```
        0      1
0  FALSE   TRUE
1  FALSE  FALSE
```

Next, we generate a corresponding data frame from `my.data` with only one absorbing state:

```
> my.data.km <- my.data
> my.data.km$to <- ifelse(my.data.km$to == "cens", "cens", 1)
```

We compute the Kaplan-Meier estimator using the `etm` package:

```
> library(etm)
> km.etm <- etm(my.data.km, c("0", "1"), tra.km, "cens", s = 0)
```

As with `mvna`, the argument `c("0", "1")` gives the state names, and `"cens"` the name marking censored observations. We also need to pass the matrix defining the possible transitions (i.e., `tra.km`). A new argument is s=0 which allows for computing *conditional* probabilities. Setting `s` to a value greater than 0 estimates $P(T > t \,|\, T > s)$. Note that $P(T > t \,|\, T > 0) = P(T > t)$. This functionality becomes more relevant with more complex multistate models.

etm returns a list: `km.etm$time` is the same as `my.fit.surv$time`. The probability estimates are in `km.etm$est`, which is an array of 2×2 matrices for each time point in `km.etm$time`. Let us write $t_1 < t_2 < t_3 < \ldots$ for the ordered observed event times contained in `km.etm$time`. The matrix that corresponds to t_1 is

```
> km.etm$est[, , 1]
```

```
        0      1
0  0.99  0.01
1  0.00  1.00
```

The upper left entry is the estimated probability of still being in the initial state by time t_1, i.e., the Kaplan-Meier estimator $\widehat{P}(T > t_1)$ evaluated at t_1. The upper right entry is the estimated probability of having left the initial state by time t_1, i.e., $\widehat{P}(T \leq t_1) = 1 - \widehat{P}(T > t_1)$. The lower left entry is the estimated probability of having made the backward transition from the absorbing state into the initial state by time t_1. This probability is 0 because $1 \to 0$ transitions are not possible. Finally, the lower right entry is the estimated probability of *not* having made the backward transition. As state 1 is absorbing, this probability is 1. Readers may check the computations by adding

```
> lines(x = km.etm$time, y = km.etm$est[1, 1, ],
+       type = "s", col = "red")
```

to Figure 4.3; the Kaplan-Meier curves should be identical.

This matrix formulation may look overdone for standard survival data, but it becomes immediately useful with more complex multistate models. E.g., imagine that we do allow for backward transitions in Figure 2.1. In biometrical applications, this model can be used for modelling healthy $\leftrightarrow$ diseased transitions, e.g., occurrence of and recovery from allergic reactions. In engineering, backward transitions might correspond to malfunctioning machines being repaired.

For convenience, probabilities may also be extracted in a simpler way using the function trprob. For instance,

```
> trprob(km.etm, "0 0")
```

displays the Kaplan-Meier estimates for all time points.

```
> trprob(km.etm, "0 0", 2)
```

will display the survival probability at time $t = 2$.

etm also estimates variances: km.etm$cov contains the estimated covariances of all probability estimators that we have just described. In the standard survival set-up, we are only interested in an estimator of the variance of the Kaplan-Meier estimator; km.etm$cov[1, 1, 1] is the Greenwood estimator (4.15) evaluated at t_1, km.etm$cov[1, 1, 2] is (4.15) evaluated at t_2, and so on. Equivalently, the trcov function may be used. trcov(km.etm, '0 0') will display the Greenwood estimator for all time points.

Cumulative incidence functions

We finally consider estimating the cumulative incidence functions $P(T \leq t, X_T = 1)$ and $P(T \leq t, X_T = 2)$. The standard way to do this in R is by using the cuminc-function of the cmprsk package:

```
> require(cmprsk)
> my.cif <- cuminc(my.data$time, my.data$to, cencode = "cens")
```

Like Surv, cuminc takes the event times my.data$time as the first argument. The second argument contains the causes of failure or a censoring code; the latter is passed to the cencode-argument. cuminc returns a list, in our example with components "1 1" and "1 2". Note that the names of these components do *not* correspond to transitions like they did in case of the mvna package. Component "1 1" contains results for failure type 1, and component "1 2" contains results for failure type 2. The first entry of these names will only differ, if we compare data from different groups defined at baseline, cf. Section 4.3. E.g., my.cif[["1 2"]]$est contains the values of $\widehat{P}(T \leq t, X_T = 2)$ at times t given in my.cif[["1 2"]]$time. Note that all *corners* of the step function $t \mapsto \widehat{P}(T \leq t, X_T = 2)$ are given. I.e., all positive times appear twice in my.cif[["1 2"]]$time. The corresponding entries in my.cif[["1 2"]]$est are $\widehat{P}(T \leq t-, X_T = 2)$ and $\widehat{P}(T \leq t, X_T = 2)$, respectively. We may plot the cumulative incidence functions via plot(my.cif). A customized output

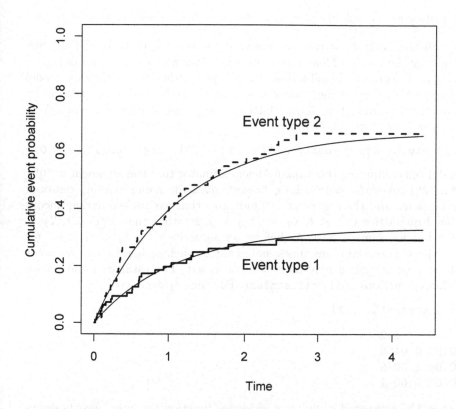

Fig. 4.4. *Simulated data.* Step functions: Aalen-Johansen estimators $\widehat{P}(T \leq t, X_T = j)$, $j = 1, 2$ for the data obs.times and obs.cause that were generated in Section 3.2. Smooth lines: theoretical cumulative incidence functions.

is given in Figure 4.4 where we have again added the theoretical quantities. Readers may check that the Aalen-Johansen estimators $\widehat{P}(T \leq t, X_T = j)$, $j = 1, 2$, add up to 1 minus the Kaplan-Meier estimator $\widehat{P}(T > t)$.

cuminc also returns an estimate of the variance of the estimator of the cumulative incidence functions. However, this variance estimator is not the Greenwood-type estimator (4.20). As the Greenwood estimator is the preferred estimator in the absence of competing risks (cf., e.g., Andersen et al., 1993, Example IV.3.1) and the performance of the variance estimator returned by cuminc has been questioned in small samples (Braun and Yuan, 2007), we do not discuss this estimator further. The estimator recommended by Braun and Yuan (2007) coincides with the Greenwood-type estimator (4.20) (Allignol et al., 2010). This estimator is provided by the etm package which we discuss next.

Estimating the cumulative incidence functions from a multistate perspective

As with the analysis of the cumulative cause-specific hazards and as with computing the Kaplan-Meier estimator using etm, we need a matrix of logical values indicating the possible transition types within the multistate model of Figure 3.1. This is the matrix tra that we used in the analysis of the cumulative hazards (cf. p. 62). Estimating the cumulative incidence functions using etm is now straightforward:

```
> cif.etm <- etm(my.data, c("0", "1", "2"), tra, "cens", s = 0)
```

Recall from computing the Kaplan-Meier estimator that the argument c("0", "1", "2") gives the state names, "cens" gives the name marking censored observations, and the argument s=0 indicates that we are estimating unconditional quantities $\widehat{P}(T \leq t, X_T = j)$, $j = 1, 2$ rather than $\widehat{P}(T \leq t, X_T = j \mid T > s)$ for some $s > 0$. As before, let us write $t_1 < t_2 < t_3 < \ldots$ for the ordered observed event times; note that at such a time t_i both an event of type 1 or of type 2 may have been observed. The estimated cumulative incidence functions are in cif.etm$est. For time t_1 we have

```
> cif.etm$est[, , 1]

      0    1 2
0 0.99 0.01 0
1 0.00 1.00 0
2 0.00 0.00 1
```

That is, the estimated cumulative incidence function for type 1 events evaluated at t_1 is $\widehat{P}(T \leq t_1, X_T = 1) = 0.01$, the top right entry is $\widehat{P}(T \leq t_1, X_T = 2) = 0$, and the top left entry is $\widehat{P}(T > t_1)$ as before. Also note that the lower two lines correspond to the lower two lines of tra containing exclusively FALSE-values (i.e., backward transitions out of the competing event states 1 and 2 are not modelled). The entries 1 on the diagonal of the lower two lines correspond to the fact that an individual stays in a competing event state, once it has been reached, with probability 1. In general, cif.etm$est evaluated at $[, , i]$ instead of $[, , 1]$ will yield

$$\widehat{P}(T > t_i) \quad \widehat{P}(T \leq t_i, X_T = 1) \quad \widehat{P}(T \leq t_i, X_T = 2)$$
$$0 \qquad\qquad 1 \qquad\qquad\qquad 0$$
$$0 \qquad\qquad 0 \qquad\qquad\qquad 1$$

In a competing risks model with potentially more than two competing event states, say k competing risks, $k \geq 2$, the first line of cif.etm$est would still display the Kaplan-Meier estimator of $P(T > t)$ in the top left entry, the next entry to the right would be the Aalen-Johansen estimator of $P(T \leq t, X_T = 1)$, and so forth; the top right entry would be the Aalen-Johansen estimator of $P(T \leq t, X_T = k)$. Also note that the Aalen-Johansen estimator, as implemented in the etm package, has a built-in guarantee that each line of the matrices in etm\$est adds up to one. For competing risks, this entails that one minus the Kaplan-Meier estimator for the waiting time in the initial state equals the sum of all estimated cumulative incidence functions. How to extract the respective estimators from the return value of etm is summarized in Table 4.1. The subscripting described in the second line of Table 4.1, using

$\widehat{P}(T > t)$	$\widehat{P}(T \leq t, X_T = 1)$	$\widehat{P}(T \leq t, X_T = 2)$
est[1,1,]	est[1,2,]	est[1,3,]
est["0","0",]	est["0","1",]	est["0","2",]
trprob(cif.etm, '00')	trprob(cif.etm, '01')	trprob(cif.etm, '02')

Table 4.1. Entries of the component **est** returned by **etm** which contains the Kaplan-Meier estimator and the Aalen-Johansen estimators of the cumulative incidence functions in a competing risks model with two competing states named '1' and '2' and one initial state named '0'. The last line displays usage of the convenience function **trprob**.

the names of the states of a multistate model, is particularly useful for more general multistate models. In Section 8.1, we will consider transition probabilities $P_{hj}(s,t) = P(X_t = j \mid X_s = h)$ for some states h, j and times $s \leq t$. With this notation, the cumulative incidence functions $P(T \leq t, X_T = j)$, $j = 1, 2$, equal transition probabilities $P_{0j}(0, t)$, and the survival function is $P_{00}(0, t)$. The indices 00 and 0j reappear in the subscripting of the last line of Table 4.1 and also in Table 4.2 below, where subscripting for an estimator of the covariance matrix is discussed.

etm also computes a Greenwood-type estimator of all variances and covariances of all transition probability estimates in a multistate model; see Section 9.2. Here, we focus on competing risks, but assessing the single components of the estimator will analogously work for general multistate models, too. The Greenwood-type estimator is contained in cif.etm\$cov. Again, cif.etm\$cov is an array, where, e.g., cif.etm\$cov[, , 1] corresponds to t_1. cif.etm\$cov[, , 1] is a matrix with dimnames

```
> dimnames(cif.etm$cov[, ,1])

[[1]]
[1] "0 0" "1 0" "2 0" "0 1" "1 1" "2 1" "0 2" "1 2" "2 2"

[[2]]
[1] "0 0" "1 0" "2 0" "0 1" "1 1" "2 1" "0 2" "1 2" "2 2"
```

That is, the rownames and the colnames are every possible combination of the states — or state names — of our multistate model. The combinations of interest to us are "0 0"; staying in the initial state 0 and "0 1" and "0 2" corresponding to transitions in either competing event state 1 or 2. The variance estimators are those entries where the rownames and the colnames are identical: cif.etm$cov["0 0","0 0",1] is the Greenwood estimator of the variance of the Kaplan-Meier estimator evaluated at t_1. cif.etm$cov["0 0","0 0",] returns all these variance estimates for $t_1 < t_2 < \ldots$. The same result may be obtained using trcov(cif.etm, '0 0'). The variance estimator for $\widehat{P}(T \leq t, X_T = 1)$ is in cif.etm$cov["0 1","0 1",], the variance estimator for $\widehat{P}(T \leq t, X_T = 2)$ is in cif.etm$cov["0 2","0 2",]. The covariance estimators are those entries, where the rownames and the colnames are not identical. E.g., cif.etm$cov["0 1","0 2",] equals cif.etm$cov["0 2","0 1",], and both these components of cif.etm$cov estimate cov$\left(\widehat{P}(T \leq t, X_T = 1), \widehat{P}(T \leq t, X_T = 2)\right)$. This mechanism is summarized in Table 4.2.

$\widehat{\text{cov}}$	$\widehat{P}(T > t)$	$\widehat{P}(T \leq t, X_T = 1)$	$\widehat{P}(T \leq t, X_T = 2)$
$\widehat{P}(T > t)$	["0 0","0 0",]	["0 0","0 1",] and ["0 1","0 0",]	["0 0","0 2",] and ["0 2","0 0",]
$\widehat{P}(T \leq t, X_T = 1)$	["0 0","0 1",] and ["0 1","0 0",]	["0 1","0 1",]	["0 1","0 2",] and ["0 2","0 1",]
$\widehat{P}(T \leq t, X_T = 2)$	["0 0","0 2",] and ["0 2","0 0",]	["0 1","0 2",] and ["0 2","0 1",]	["0 2","0 2",]

Table 4.2. etm returns an array as a value cov, which contains Greenwood-type estimators of the covariances of the Kaplan-Meier estimator $\widehat{P}(T > t)$ and the estimators $\widehat{P}(T \leq t, X_T = j)$ of the cumulative incidence functions. The table denotes the respective entries of cov where these covariances are found.

The value returned by etm allows for straightforward computation of the pointwise confidence intervals (4.21) of $\widehat{P}(T \leq t, X_T = j)$. For computing 95% confidence intervals for $\widehat{P}(T \leq t, X_T = 1)$, the argument (except for the sign) of exp in (4.21) is

```
qnorm(0.975) * sqrt(cif.etm$cov["0 1","0 1",])/
((1-cif.etm$est[1,2,]) * log(1-cif.etm$est[1,2,]))
```

We consider confidence intervals in the analysis of the pneumonia data below (cf. Figure 4.10). For convenience, confidence intervals may also be obtained using the summary function.

Left-truncation

Our presentation of nonparametric estimation has so far focused on tools that are available in R and their ability to recover the theoretical quantities of the competing risks multistate model in the presence of right-censoring. We now illustrate that this program also works in the presence of left-truncation. Recall from Section 2.2.2 that left-truncation arises in situations of delayed study entry: individuals are not under observation since time origin, but only enter the study at some later point in time. They enter the risk set at their time of study entry and may be observed to experience a competing event only *after* their entry time. Only individuals enter the study whose time of study entry is less than their event time and less than their censoring time (cf. (4.1)).

We additionally generate left-truncation times,

```
> lt.times <- rgamma(100, shape= 0.5, rate = 2)
```

Left-truncation is light, the number of individuals under study (out of 100 individuals originally simulated) is

```
> sum(lt.times < my.data$time)
```

```
[1] 82
```

The cumulative cause-specific hazards may again be conveniently computed using mvna. We only need to slightly recode the data in order to inform mvna about the entry times into the risk set and about the exit times.

```
> my.data2 <- my.data[,1:3] #do not keep my.data$time
> my.data2$entry <- lt.times
> my.data2$exit <- my.data$time
> my.data2 <- my.data2[my.data2$entry<my.data2$exit,]
> head(my.data2)
```

	id	from	to	entry	exit
1	1	0	2	0.14158852	0.1450839
2	2	0	2	0.60343943	2.3988344
3	3	0	1	0.34543187	0.4447219
5	5	0	1	0.04906455	0.6050701
6	6	0	2	0.18248723	0.8025029
8	8	0	2	0.07326980	0.2346617

Instead of one entry time, my.data2 now has entries entry and exit. entry contains the left-truncation times. exit contains the minimum of an individual's event time and right-censoring time. Only individuals with left-truncation times less than the times in exit enter the study.

We can now use mvna as before, working, however, with the left-truncated data set my.data2,

```
> my.nelaal2 <- mvna(my.data2, c("0", "1", "2"), tra, "cens")
```

The Nelson-Aalen estimates for both possible transitions along with the log-transformed confidence intervals of (4.10) may again be plotted using xyplot,

```
> xyplot(my.nelaal2, strip=strip.custom(bg="white"),
+        ylab="Cumulative Hazard", lwd=2)
```

The output is displayed in Figure 4.5 where we have again included the theoretical quantities $A_{01}(t) = 0.3t$ and $A_{02}(t) = 0.6t$ as in Figure 4.2. The

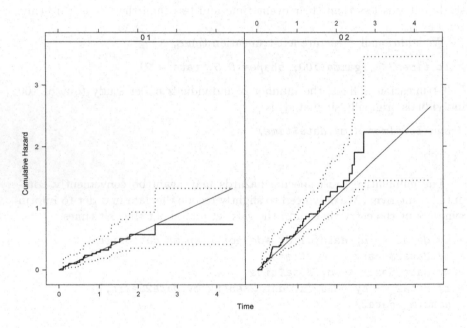

Fig. 4.5. *Simulated data.* The step functions are the Nelson-Aalen estimates $\widehat{A}_{01}(t)$ (left) and $\widehat{A}_{02}(t)$ (right) with log-transformed confidence intervals for the data as in Figure 4.2, but now additionally subject to left-truncation. The linear functions are the true cumulative hazards $A_{01}(t)$ and $A_{02}(t)$.

approximation of the true cumulative hazards by the Nelson-Aalen estimates is still satisfactory, but the confidence intervals are now wider for early times as compared to Figure 4.2. This is a consequence of left-truncation, which typically leads to smaller risk sets, in particular for early points in time. This is illustrated in Figure 4.6.

Recalling that the Kaplan-Meier estimator $\widehat{P}(T > t)$ and the Aalen-Johansen estimators $\widehat{P}(T \leq t, X_T = j)$ are deterministic functions of the

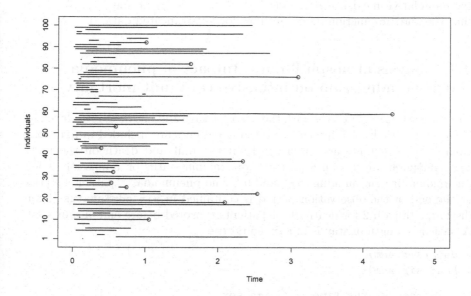

Fig. 4.6. *Simulated data.* The solid lines indicate the individual times at risk for the simulated data and in the presence of left-truncation. Right-censored individuals are indicated with a circle o. Note that the plot shows less than 100 lines, because some individuals do not enter the study as a consequence of left-truncation.

Nelson-Aalen estimators computed above, we note that these probability estimates may also be computed in the presence of left-truncation, but that we have to reckon with wider confidence intervals for small t. We briefly look at the necessary changes which need to be made in the R code, and a further discussion follows in the real data examples below.

$\widehat{P}(T > t)$ may again be computed using the surv fit-function,

```
> my.fit.surv2 <- survfit(Surv(entry,exit, to != "cens") ~ 1,
+                         data = my.data2, conf.type = "log-log")
```

As with mvna, the difference is that we now pass the entry times entry into the study cohort and the exit times exit out of the study cohort to survfit. In the context of the survival package, this is also often referred to as start stop-coding or 'counting process data'.

Cumulative incidence functions may conveniently be estimated using etm. The required changes in R are as with mvna; we use the data set my.data2.

```
> cif.etm2 <- etm(my.data2, c("0", "1", "2"),
+                 tra, "cens", s = 0)
```

Recall that the use of etm as in the above code estimates both P($T > t$) and the cumulative incidence functions P($T \leq t, X_T = j$) in one step. We note that the cuminc-function does not allow for left-truncated data.

4.3 Analysis of hospital data: Impact of pneumonia status on admission on intensive care unit mortality

The data set sir.adm comes with the mvna package and has been introduced in Chapter 1. Briefly, 747 intensive care unit patients are included in sir.adm. Competing endpoints are discharge from the unit and death on the unit. sir.adm$pneu informs on a patient's pneumonia status on admission, 1 for pneumonia present on admission, and 0 for no pneumonia. A patient's status at the end of the observation period is contained in sir.adm$status, 1 for discharge (alive), 2 for death, 0 for patients censored before end of unit stay. A patient's length of stay is in sir.adm$time.

```
> data(sir.adm)
> head(sir.adm)

   id pneu status time      age sex
1   41    0      1    4 75.34153   F
2  395    0      1   24 19.17380   M
3  710    1      1   37 61.56568   M
4 3138    0      1    8 57.88038   F
5 3154    0      1    3 39.00639   M
6 3178    0      1   24 70.27762   M
```

The aim of the present analysis is to study the impact of pneumonia present on admission on unit mortality. As pneumonia is a severe illness, we should expect more patients dying with pneumonia than without. In terms of Figure 3.1, death is the event of interest and discharge is the competing event.

We first study the cumulative cause-specific hazards using mvna. We use the same matrix tra of logical values indicating the possible transition types within the competing risks multistate model as before. As in the analysis of the simulated data, we need to modify sir.adm into a multistate-type data set. We also recode the data such that the event of interest, death, will correspond to state 1 as in Figure 3.1, and state 2 is the competing event:

```
> to <- ifelse(sir.adm$status==0,"cens",
+              ifelse(sir.adm$status==1,2,1))
> my.sir.data <- data.frame(id=sir.adm$id,from=0,to,
+                           time=sir.adm$time, pneu=sir.adm$pneu)
```

Note that my.sir.data has a component pneu with the pneumonia on admission status. We check the recoding of the nested call to ifelse,

```
> table(my.sir.data$to)
```

```
    1     2 cens
   76   657    14
```

Next, we compute the cumulative cause-specific hazards for death and discharge, respectively, and stratified for pneumonia status on admission,

```
> my.nelaal.nop <- mvna(my.sir.data[my.sir.data$pneu == 0, ],
+                       c("0", "1", "2"), tra, "cens")
> my.nelaal.p <- mvna(my.sir.data[my.sir.data$pneu == 1, ],
+                       c("0", "1", "2"), tra, "cens")
```

A customized plot of the Nelson-Aalen estimates is displayed in Figure 4.7. Note that pneumonia appears to have no effect on the death hazard. However,

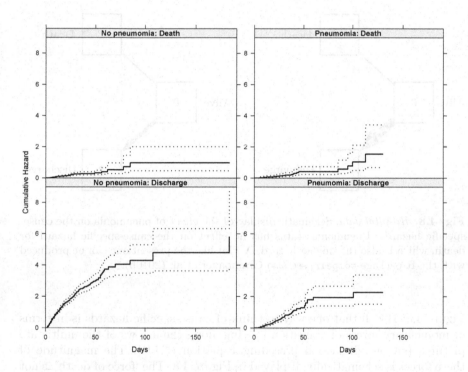

Fig. 4.7. *Hospital data.* Top row: Nelson-Aalen estimates $\widehat{A}_{01}(t)$ of the cumulative death hazard. Bottom row: Nelson-Aalen estimates $\widehat{A}_{02}(t)$ of the cumulative discharge hazard. All estimates are stratified for pneumonia status on admission.

this does *not* imply that pneumonia has no effect on mortality. The reason is that pneumonia appears to reduce the discharge hazard. This implies:

1. Pneumonia appears to reduce the all-cause hazard for end of intensive care unit stay.

2. Patients with pneumonia on admission stay longer on the unit. During this prolonged stay, they are exposed to an essentially unchanged death hazard.
3. As a consequence, more patients with pneumonia die than patients without pneumonia.

This is a typical competing risks phenomenon. Because there is more than one hazard acting on an individual, we cannot tell from one hazard alone what an individual's future course will be. This situation is schematically displayed in

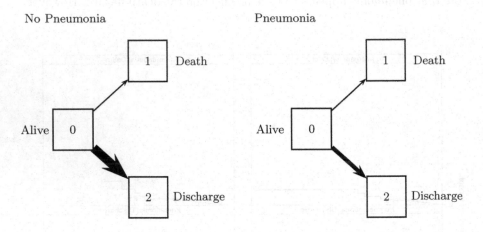

Fig. 4.8. *Hospital data.* Schematic display of the effect of pneumonia on the cause-specific hazards. Pneumonia status has no effect on the cause-specific hazard for death, which is also the smaller hazard. A Plot like the present one can be produced with the R package `compeir`; see also Grambauer et al. (2010b).

Figure 4.8: Recall that one way to think of cause-specific hazards is in terms of momentary forces of transition moving along the arrows of the multistate pictures (cf. our discussion preceding Equation (3.14)). The magnitude of these forces is schematically displayed in Figure 4.8. The 'force of death' is not influenced by pneumonia status, but the 'force of discharge' is substantially reduced by pneumonia on admission. Figure 4.8 illustrates that the 'overall force', i.e., the all-cause hazard that pulls an individual is reduced, leading to longer unit stay, and that the relative strength between the cause-specific forces of death and of discharge is changed by pneumonia status.

Note that the schematic representation of Figure 4.8 has limitations. The magnitude of the momentary transition forces will usually not be constant over time, such that we would actually need a whole series of plots like Figure 4.8. In fact, this is achieved in Figure 4.7: the *shape* of the Nelson-Aalen estimators, which estimate the *cumulative* hazards, is determined by the cause-specific hazards. We may also think of Figure 4.8 in a such way that does not necessarily illustrate the magnitude of the hazards, which may vary with time, but only the ratios of the death hazards and the ratios of the discharge hazards, respectively, which are assumed to be constant. This is the approach taken by proportional cause-specific hazards modelling which we consider in Section 5.

Finally, we check whether our interpretation of the cumulative hazards analysis has been correct by looking at the Aalen-Johansen estimators of the cumulative incidence functions, again stratified for pneumonia status. Recall that the cumulative incidence function for death, say, displays the expected proportion of individuals dying on the unit over the course of time. If our interpretation of the cumulative hazards analysis has been correct, the estimated cumulative incidence function for death, $\widehat{P}(T \leq t, X_T = 1)$, within patients with pneumonia should run above those patients without pneumonia.

Using the function `cuminc` of the `cmprsk` package, we may now also pass a `group = my.sir.data$pneu` argument in the call to `cuminc`. This forces `cuminc` to compute the estimates $\widehat{P}(T \leq t, X_T = j)$, $j = 1, 2$ within groups defined by `my.sir.data$pneu`.

```
> my.sir.cif <- cuminc(my.sir.data$time, my.sir.data$to,
+                       group=my.sir.data$pneu, cencode="cens")
```

The return value of `cuminc` is a list with components "0 1", "1 1", "0 2" and "1 2". The components "0 1" and "1 1" contain results for failure type 1; the components "0 2" and "1 2" contain results for failure type 2. The components "0 1" and "0 2" are for patients with pneumonia status 0 on admission, i.e., without pneumonia, and components "1 1" and "1 2" are for patients with pneumonia status 1. A customized plot of the estimated cumulative incidence functions is displayed in Figure 4.9.

As expected, we find that more patients die among those with pneumonia.

The computations may also be done using the `etm` package. As with `mvna`, we run `etm` with each stratum:

```
> my.sir.etm.nop <- etm(my.sir.data[my.sir.data$pneu == 0, ],
+                        c("0", "1", "2"), tra, "cens", s = 0)
> my.sir.etm.p <- etm(my.sir.data[my.sir.data$pneu == 1, ],
+                      c("0", "1", "2"), tra, "cens", s = 0)
```

The Aalen-Johansen estimates $\widehat{P}(T \leq t, X_T = 1)$ are shown in Figure 4.10, generated by

```
> plot(my.sir.etm.p, tr.choice = '0 1', col = 1, lwd = 2,
+       conf.int = TRUE, ci.fun = "cloglog", legend = FALSE,
+       ylab="Probability")
> lines(my.sir.etm.nop, tr.choice = '0 1', col = "gray",
+        lwd = 2, conf.int = TRUE, ci.fun = "cloglog")
```

together with pointwise 95% confidence intervals (4.21).

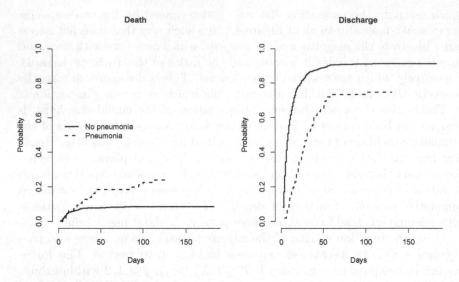

Fig. 4.9. *Hospital data.* Aalen-Johansen estimates $\widehat{P}(T \leq t, X_T = j)$ of the cumulative incidence functions for death (left, $j = 1$) and for discharge (right, $j = 2$), stratified for pneumonia status on admission. Solid lines are for patients without pneumonia.

The confidence intervals support our previous finding that we eventually see more cases of death in the group of pneumonia patients.

Plots of cumulative hazards as in Figure 4.7 and plots of cumulative incidence functions as in Figures 4.9 and 4.10 both have their relative merits: obviously, it is easier to tell from Figure 4.9 whether pneumonia increases unit mortality. However, we need to look at the cumulative cause-specific hazards to see whether increased mortality is due to an increase of the death hazard, say, or — as in the present example — due to a decrease of the discharge hazard. We return to this tradeoff in Section 5 where we discuss regression modelling of the cause-specific hazards and direct regression modelling of the cumulative incidence functions. In Section 5, we also discuss examples, where we find a unidirectional effect on all of the cause-specific hazards. This usually leads to crossing cumulative incidence functions.

4.3.1 Latent failure time approach

We briefly return to the latent failure time approach towards competing risks; see Section 3.3. The multistate approach taken thus far has assumed that a patient has a time of hospital stay T, and that at time T (i.e., at the end of hospital stay), there is also a vital status X_T, either 'hospital death' or 'alive discharge'. In contrast, assuming latent failure times posits the existence of

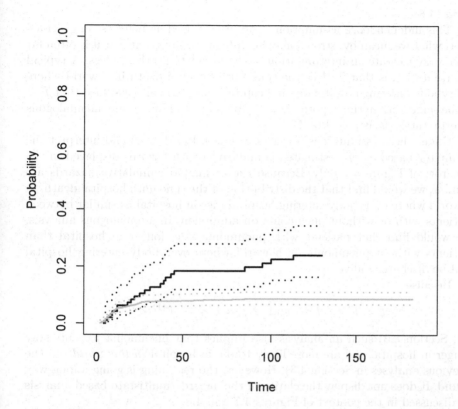

Fig. 4.10. *Hospital data.* The solid lines are the Aalen-Johansen estimates $\widehat{P}(T \leq t, X_T = 1)$ for event type 'death' and stratified for pneumonia status on admission. The black lines are for pneumonia patients, the grey lines are for patients without pneumonia. The dashed lines indicate the pointwise 95% confidence intervals of Equation (4.21).

times $T^{(1)}, T^{(2)}$, where both the time of hospital stay and the vital status at the end of hospital stay are determined by the smaller of the two latent times $T^{(1)}$ and $T^{(2)}$.

The latent failure time model is typically being criticized, because any dependence structure between $T^{(1)}$ and $T^{(2)}$ cannot be verified from the data, because the physical interpretation of $T^{(1)}$ and $T^{(2)}$ is awkward and because assuming latent times does not constitute a fruitful approach to answer questions of the original subject matter, hospital epidemiology in this case.

We briefly illustrate these issues based on the previous data analyses. To begin, we assume that $T^{(1)}$ and $T^{(2)}$ are, in fact, independent. This assumption is often made for reasons of tractability: under independence, the Kaplan-Meier-type estimator $\prod_0^t \left(1 - \mathrm{d}\widehat{A}_{01}(u) \right)$ estimates $\mathrm{P}(T^{(1)} > t)$, where $\widehat{A}_{01}$ is as in (4.8).

This independence assumption is often criticized as being 'strong'. However, what we mean by 'strong' may be difficult to say given that it is often far from easy to state an interpretation of the latent times themselves. A typical interpretation is that $T^{(1)}$ is 'the time until hospital death in a world where everybody entering hospital dies in hospital'. It is then also assumed that $T^{(1)}$ is the same both in the hypothetical world and in real life. Again, an analogous interpretation is attached to $T^{(2)}$.

Under these assumptions, we may estimate $\mathrm{P}(T^{(1)} > t)$ and interpret the estimator based on the estimated cumulative death hazards displayed in the top row of Figure 4.7 only. Because the estimated cumulative hazards are similar, we would find that the distribution of the time until hospital death, in a world where everybody entering hospital dies in hospital, is similar between patients with or without pneumonia on admission. In an analogous analysis, we would find that patients with pneumonia stay longer in hospital than patients without pneumonia, in a world where everybody entering hospital will be discharged alive.

Because

$$T = T^{(1)} \wedge T^{(2)} \ \text{ and } \ X_T = 1 \iff T^{(1)} < T^{(2)}$$

(cf. Section 3.3) such an analysis also implies that pneumonia patients stay longer in hospital and are more likely to die in hospital *in this world* (cf. the previous analyses in Section 4.3). However, the reasoning is going a long way round. It does not display the clarity of the hazard-/multistate-based analysis as discussed in the context of Figures 4.7 and 4.8.

One attraction towards latent failure time modelling lies in speculation on hospital stay and hospital outcome under hypothetical modifications of the competing events, including cause removal. As explained in Section 3.3, such hypothetical calculations can easily be done without assuming latent times. Instead, calculations would be based on modifying the cause-specific hazards displayed in Figure 4.7.

4.4 Analysis of pregnancy outcome data: Impact of coumarin derivatives on spontaneous and induced abortion

We now illustrate how to estimate cumulative incidence functions in the presence of left-truncation. The data set `abortion` comes with the `etm` package and has been introduced in Chapter 1. Briefly, 1186 women are included in the data set. Competing endpoints are spontaneous abortion, induced abortion, and live birth. Women therapeutically exposed to coumarin derivatives have value 1 in `abortion$group`, which is 0 otherwise. Pregnancy outcomes are in `abortion$cause`, 1 for induced abortion, 2 for live birth and 3 for spontaneous abortion. The data are left-truncated: time origin is conception, but women do not enter the study before the pregnancy is recognized. Left-truncation times are in `abortion$entry`, times of live birth or abortion are in `abortion$exit`. Right-censoring did not occur.

```
> data(abortion)
> head(abortion)

  id entry exit group cause
1 1     6   37     0     2
2 2     9   40     0     2
3 3    29   40     0     2
4 4    32   41     0     2
5 5    11   39     0     2
6 6    10   39     0     2
```

The empirical distribution of the left-truncation times, i.e., the times of study entry, is displayed in Figure 4.11. The curves in Figure 4.11 estimate the

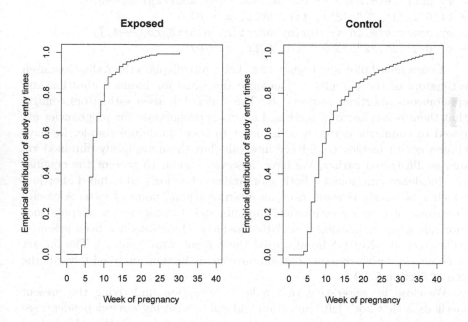

Fig. 4.11. *Pregnancy outcome data.* Empirical distribution of the observed study entry times.

distribution of the study entry time *given study entry*, i.e., given $L < T$. The figure illustrates that delayed study entry is an issue in this data set.

The aim of the present analysis is to study the impact of coumarin derivatives on the abortion proportions, which are suspected to be increased by the medication.

For convenience, we split the data into women exposed to coumarin derivatives and the control group, and we slightly modify the data structure for use with etm.

```
> my.abort <- abortion
> my.abort$from <- rep(0,nrow(my.abort))
> names(my.abort)[5] <- c("to") #rename cause
```

Unlike the previous examples, we now have three competing events. This is reflected in tra,

```
> tra <- matrix(FALSE, nrow = 4, ncol = 4)
> tra[1, 2:4] <- TRUE
```

Interest focuses on abortion, either spontaneous or induced, but we have chosen to keep the state names as in abortion$cause. We compute the Aalen-Johansen estimator within the two groups:

```
> my.abort.etm.nocd <- etm(my.abort[my.abort$group==0,],
+ c("0", "1","2","3"), tra, NULL, s = 0)
> my.abort.etm.cd <- etm(my.abort[my.abort$group==1,],
+ c("0", "1","2","3"), tra, NULL, s = 0)
```

A customized plot is in Figure 4.12. The figure displays the Aalen-Johansen estimators of the cumulative incidence functions for induced abortion and spontaneous abortion, respectively. The Aalen-Johansen estimators confirm that there is a concern of increased abortion proportions for pregnancies exposed to coumarin derivatives. In order to keep the figure simple, we have chosen not to include confidence intervals, but these are easily obtained via etm as illustrated earlier. We have, however, chosen to present the cumulative incidence functions for both spontaneous abortion and induced abortion, which is obviously relevant from an interpretational point of view. A careful discussion of variance estimation and subsequent construction of confidence intervals when estimating cumulative incidence functions has been given by Allignol et al. (2010), who also used the present data example. The impact of coumarin derivatives on abortion outcome is further analysed towards the end of Section 5.2.2.

We close by reiterating that a key assumption underlying the present methods is that both left-truncation and right-censoring are independent; see Section 2.2.2. In the present example, one may wonder whether this is true with respect to the outcome 'induced abortion'. Unlike independent right-censoring, the assumption of independent left-truncation may be empirically investigated; we discuss this issue in Section 11.3. For the time being, we note that this assumption turned out to be justifiable for the present data.

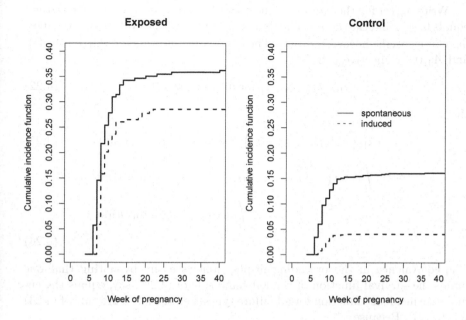

Fig. 4.12. *Pregnancy outcome data.* Estimated cumulative incidence functions of abortion. The left plot is for women exposed to coumarin derivatives. The dashed lines indicate induced abortion; the solid lines indicate spontaneous abortion.

4.5 A special case: No effect on the competing cause-specific hazard

We have characterized interpretational challenges as in the pneumonia example of Section 4.3 as a 'typical competing risks phenomenon'. A careful discussion of interpretational subtleties that are even more involved is given in Section 5.2.2. There are, however, cases where the interpretation of competing risks data is straightforward: such an interpretationally more convenient situation arises if a covariate displays an effect on the cause-specific hazard of interest only and has no effect on the competing hazard. This situation has, e.g., received some attention in sample size planning for competing risks data (Schulgen et al., 2005; Latouche and Porcher, 2007).

In such a situation, a reducing effect of a covariate on $\alpha_{01}(t)$ with no effect on $\alpha_{02}(t)$ will lead to a reduced cumulative incidence function for event type 1 for all times $t > 0$. In Section 5.2.2, we find that judging the impact of a covariate in terms of the cumulative event probabilities will usually require us to carefully consider the magnitude of the cause-specific hazards, too. In the present situation, things are simpler, and we give a brief proof of this important fact.

Write $\alpha_{01}(t)$ for the cause-specific hazard of interest, if a binary covariate equals 0, say, and $\widetilde{\alpha}_{01}(t)$, if it equals 1. For both covariate values, the competing cause-specific hazard is $\alpha_{02}(t)$. The cumulative hazards are $A_{01}(t)$, $\widetilde{A}_{01}(t)$, and $A_{02}(t)$, respectively. If

$$\widetilde{\alpha}_{01}(t) < \alpha_{01}(t) \text{ for all times } t > 0, \tag{4.22}$$

then

$$\exp\left(-\widetilde{A}_{01}(t) - A_{02}(t)\right) > \exp\left(-A_{01}(t) - A_{02}(t)\right) \tag{4.23}$$

$$\Longleftrightarrow \int_0^t \exp\left(-\widetilde{A}_{01}(u) - A_{02}(u)\right)(\widetilde{\alpha}_{01}(u) + \alpha_{02}(u))\mathrm{d}u$$

$$< \int_0^t \exp\left(-A_{0\cdot}(u)\right)\alpha_{0\cdot}(u)\mathrm{d}u. \tag{4.24}$$

The equivalence in the preceding display follows easily by noting that one minus the survival function of the left hand side of (4.23), say, equals the cumulative incidence function for all failure types, i.e., the left hand side of (4.24) (cf. (3.12)). Because

$$\int_0^t e^{-\widetilde{A}_{01}(u) - A_{02}(u)}\alpha_{02}(u)\mathrm{d}u > \int_0^t e^{-A_{01}(u) - A_{02}(u)}\alpha_{02}(u)\mathrm{d}u,$$

Equation (4.24) implies that

$$\int_0^t e^{-\widetilde{A}_{01}(u) - A_{02}(u)}\widetilde{\alpha}_{01}(u)\mathrm{d}u < \int_0^t e^{-A_{01}(u) - A_{02}(u)}\alpha_{01}(u)\mathrm{d}u \tag{4.25}$$

for all times $t > 0$, i.e., the cumulative incidence function for event type 1 is reduced, as claimed earlier. Of course, these inequalities are reversed, if $\widetilde{\alpha}_{01}(t) > \alpha_{01}(t)$, leading to an increased cumulative incidence function.

Recalling our discussion of Figure 4.8 where we interpreted the cause-specific hazards as momentary forces of transition moving along the arrows of a multistate picture, the inequality in (4.25) is an intuitive result. To illustrate this, we briefly reconsider the pneumonia example of Section 4.3:

Assume that 'alive discharge' were the event of interest instead of 'hospital death'. As pneumonia is a severe illness, we should expect *fewer* patients being discharged alive with pneumonia than without. In fact, the analysis in Section 4.3 finds a reduced cumulative incidence function for alive discharge, because the cause-specific hazard for discharge is reduced by pneumonia with no effect on the cause-specific hazard for death. Note that this result has essentially been our interpretation of the Nelson-Aalen estimates in Figure 4.7.

We note that it is usually a question of the subject matter research whether a competing event state is interpreted as the 'event of interest'. In the pneumonia example, this clearly is 'hospital death'. Hence, potential difficulties when interpreting or communicating results such as 'no effect on the cause-specific hazard for death, but an increase in hospital mortality' should not be tackled by redefining the 'event of interest'. In addition, Section 5.2.2 gives an example where such an approach would yield no benefit whatsoever. However, both examples illustrate that any competing risks analysis will remain incomplete, does not allow for a probability interpretation, and may potentially be misleading unless all cause-specific hazards are analysed.

Finally, we note that the cumulative incidence function for type 1 events is also reduced for all times $t > 0$, if, in addition to $\widetilde{\alpha}_{01}(t) < \alpha_{01}(t)$, there is an opposite effect on the competing cause-specific hazards, i.e., $\widetilde{\alpha}_{02}(t) > \alpha_{02}(t)$. The result easily follows from the notion of the cause-specific hazards as momentary forces of transition or by noting that $\int_0^t \exp(-\widetilde{A}_{0.}(u))\widetilde{\alpha}_{01}(u)\mathrm{d}u$ is less than the left hand side of (4.25).

4.6 Exercises

1. Show that the 'cumulative incidence' representation of the Kaplan-Meier estimator, holds, i.e., verify Equation (4.17).
2. Show that the Aalen-Johansen estimator of $P(T \leq t, X_T = 1)$ is always less than or equal to one minus $\prod_{T_i \wedge C_i \leq t} \left(1 - \frac{\Delta N_{01}(T_i \wedge C_i)}{Y_0(T_i \wedge C_i)}\right)$.
3. *The 4D study*: The data set fourD contains the data from the 4D study for the *placebo* group. The data set is part of the etm package and has been described in Chapter 1. Briefly, fourD\$status describes a patient's status at the end of the follow-up, 1 for the event of interest, 2 for death from other causes, and 0 for censored observations. fourD\$time contains the follow-up time.

   a) Estimate the cumulative cause-specific hazards in the original placebo data using the mvna package. Compute the Kaplan-Meier estimator, starting from the increments of both Nelson-Aalen estimates. Also estimate the survival function of the censoring times.

   b) Simulate *new* placebo group data for 500 individuals using the empirical distributions of the previous step. (Hint: The last observed follow-up time is censored. As a consequence, the Kaplan-Meier estimator does not spent 100% of the probability mass. Place the remaining probability mass at a time point beyond the largest follow-up time. If this time point is sampled in the simulation, the corresponding follow-up time will always be censored.)

   c) Simulate treatment group data for 500 individuals. Assume that the ratio *treatment vs. placebo* of the cause-specific hazard of interest is $\exp(-0.1)$. Assume that the competing cause-specific hazard is the same for both treatment and placebo. Use the censoring distribution as in the placebo group.

    d) Perform the nonparametric analyses of this chapter using the simulated data (i.e., Nelson-Aalen, Kaplan-Meier, Aalen-Johansen).

4. Consider the situation of Exercise 3. Explain the different constellations of the cumulative incidence functions after manipulation of the cause-specific hazards as in the following table.

Scenario	β_1	β_2	Transformation of $\widehat{A}_{01}(t)$	Transformation of $\widehat{A}_{02}(t)$
1	-0.3	0	none	none
2	-0.3	0	$x \mapsto x^{1/4}$	none
3	-0.3	0.3	none	none
4	-0.3	0.3	none	$x \mapsto x^2$
5	-0.3	-0.3	$x \mapsto x^2$	$x \mapsto x^{1/4}$

In the table, $\widehat{A}_{01}(t)$ is the Nelson-Aalen estimator of the cumulative cause-specific hazard of interest in the placebo group. The corresponding estimator of the competing cumulative cause-specific hazard is $\widehat{A}_{02}(t)$. The log of the ratio of the cause-specific hazards of interest is β_1. The corresponding value for the competing cause-specific hazards is β_2.

5. Add left-truncation to the data simulated in Exercise 3 and redo the analyses. Draw a plot as in Figure 4.6. Left-truncation times shall be drawn from a log-normal distribution (function `rlnorm`) with parameters chosen to let approximately 70% of the individuals enter the study.

6. Repeat the analyses of the pregnancy outcome data from a latent failure time perspective as in Section 4.3.1.

5

Proportional hazards models

This chapter discusses the most widely used regression models in competing risks. Following an introduction in Section 5.1, Section 5.2 discusses proportional cause-specific hazards models, and Section 5.3 discusses the proportional subdistribution hazards model. The cause-specific hazards are as defined in Chapter 3. The subdistribution hazard is a different hazard notion, namely the hazard 'attached' to the cumulative incidence function of interest as explained below. Both modelling approaches have their relative merits, and both approaches make the proportional hazards assumption solely for interpretational and technical convenience. It is not uncommon that both approaches are employed in one data analysis, although one model assumption usually precludes the other. A justification for employing both models side-by-side is provided in Section 5.4. Goodness-of-fit methods are described in Section 5.5, and Section 5.6 gives a brief overview of regression models that go beyond the familiar proportional hazards assumption together with their availability in R.

As in Chapter 4, we consider n individuals under study with competing risks data subject to right-censoring and left-truncation. The analysis of cause-specific hazards in Section 5.2 assumes right-censoring and left-truncation to be independent (cf. Section 2.2.2). These restrictions on the observable competing risks data may depend on the past. In practice, this entails that right-censoring and left-truncation may depend on covariates included in the model. In contrast, the analysis using subdistribution hazards in Section 5.3 assumes that the competing risks data are subject to random censoring only. In particular, right-censoring may not depend on covariates. The reason for the more restrictive assumption of random censorship is a certain technical difficulty encountered in the subdistribution framework. This is explained in detail in Section 5.3.2.

5.1 Introduction

So far, we have considered competing risks data arising from homogeneous groups. This means that, safe for right-censoring and left-truncation, and after potential stratification as described in Section 4.3, the data have been considered to be independent copies of a competing risks process with cause-specific hazards $\alpha_{0j}(t)$, $j = 1, 2$. In Section 5.2, we study proportional cause-specific hazards models, relating the α_{0j}s to a vector of covariates Z_i for individual i, $i = 1, \ldots, n$, known at time origin, that is baseline covariates such as gender and age. We assume that the n competing risks processes are conditionally independent given the baseline covariate values. A hazard regression model can also be formulated for time-dependent covariates, but the interpretation becomes more difficult. Time-dependent covariates are better dealt with from a more general multistate perspective, and we do so in Chapter 11.

Proportional cause-specific hazards models assume each cause-specific hazard to follow a Cox model (Cox, 1972)

$$\alpha_{0j;i}(t; Z_i) = \alpha_{0j;0}(t) \cdot \exp\left(\beta_{0j} \cdot Z_i\right), \, j = 1, 2, \, i = 1, \ldots, n, \qquad (5.1)$$

where β_{0j} is a $1 \times p$ vector of regression coefficients, Z_i is a $p \times 1$ vector of covariates for individual i, and $\alpha_{0j;0}(t)$ is an unspecified, non-negative baseline hazard function. We also write

$$A_{0j;0}(t) = \int_0^t \alpha_{0j;0}(u)\mathrm{d}u \quad \text{and} \quad A_{0j;i}(t; Z_i) = \int_0^t \alpha_{0j;i}(u; Z_i)\mathrm{d}u \qquad (5.2)$$

for the respective cumulative cause-specific hazards, $j = 1, 2$, $i = 1, \ldots, n$.

The two models (5.1) for $j = 1, 2$ are semiparametric in the sense that the baseline hazard function is an element of an infinite-dimensional function space, and the vector of regression coefficients is an element of $\mathbb{R}^p$. They specify that the cause-specific hazard ratios between two individuals with, say, observed covariate vectors z_1 and z_2 are

$$\exp\left(\beta_{0j} \cdot z_1 - \beta_{0j} \cdot z_2\right), \, j = 1, 2.$$

If $z_1 = (z_{11}, \ldots, z_{1p})^\top$ equals $z_2 = (z_{21}, \ldots, z_{2p})^\top$ except for $z_{11} = z_{21} + 1$, this implies that the cause-specific hazard $\alpha_{0j;1}(t)$ of individual 1 is $\exp(\beta_{0j1})$ times the cause-specific hazard of individual 2 where $\beta_{0j} = (\beta_{0j1}, \ldots, \beta_{0jp})$. Analogously, $\exp(\beta_{0jk})$ reflects an one-unit increase in the kth entry of the covariate vector, and all other covariates are kept fixed.

We have formulated models (5.1) with cause-specific vectors β_{0j} of regression coefficients. Sometimes, one may wish to model a common effect of a covariate on both cause-specific hazards. It is therefore useful to reformulate (5.1) with only one vector β, which contains all regression coefficients and does not depend on the event type, and cause-specific vectors $Z_{0j;i}$ of covariates:

$$\alpha_{0j;i}(t; Z_i) = \alpha_{0j;0}(t) \cdot \exp\left(\beta \cdot Z_{0j;i}\right), \, j = 1, 2, \, i = 1, \ldots, n. \tag{5.3}$$

This reformulation is of direct importance for R programming; see the analysis of the data frame xl in Section 5.2.2. How to move from (5.1) to (5.3) can easily be demonstrated in examples.

Assume that Z_i is a real-valued covariate with cause-specific regression coefficients β_{01} and β_{02}. We combine the regression coefficients into one vector $\beta = (\beta_{01}, \beta_{02})$. The cause-specific covariate vectors are $Z_{01;i} = (Z_i, 0)^\top$ and $Z_{02;i} = (0, Z_i)^\top$. Next, we assume that we have two covariates for each individual, $Z_i = (Z_{i1}, Z_{i2})$ and that Z_{i1} has cause-specific regression coefficients β_{0j1} for event type j, $j = 1, 2$, but that Z_{i2} displays a common effect $\beta_{0j2} = \eta$. We combine the different regression coefficients into one vector $\beta = (\beta_{011}, \beta_{021}, \eta)$. The cause-specific covariate vectors are $Z_{01;i} = (Z_{i1}, 0, Z_{i2})^\top$ and $Z_{02;i} = (0, Z_{i1}, Z_{i2})^\top$.

Proportional cause-specific hazards models are the standard regression technique in competing risks. In the absence of a competing hazard, the models reduce to a standard Cox survival model. Interpretation in terms of the survival probability is then straightforward. If the outcome is death, a death hazard ratio less than 1 is beneficial, a hazard ratio greater than 1 is harmful. The interpretation becomes more involved with competing risks: recall from the analysis of the pneumonia data in Section 4.3 that more patients with pneumonia die in hospital than patients without pneumonia. However, this increase is not the result of an increased cause-specific hazard for hospital death, but of a decreased cause-specific hazard for alive discharge. The cause-specific hazard for hospital death is left unchanged by pneumonia status. We revisit this data example in Section 5.2.2 where we confirm our previous findings using proportional cause-specific hazards models.

Interpretational difficulties like these — 'increase in mortality, but no effect on the death (cause-specific) hazard' — have led to modelling subdistribution hazards. A subdistribution hazard analysis offers a synthesis of the cause-specific hazards analyses. The effects that a baseline covariate displays on the cause-specific hazards are summarized in terms of an effect that the covariate displays on one cumulative incidence function. The key idea is to introduce a new hazard notion, the subdistribution hazard, for the event of interest (such as hospital death), which reestablishes a one-to-one correspondence with the cumulative incidence function. Recall from (3.11) that the cumulative incidence function for event type 1, say, is an involved function of all cause-specific hazards:

$$P(T \leq t, X_T = 1) = \int_0^t \exp\left(-\int_0^u \alpha_{01}(v) + \alpha_{02}(v)dv\right) \alpha_{01}(u) \, du.$$

The subdistribution hazard $\lambda(t)$ is required to fulfill

$$P(T \leq t, X_T = 1) = 1 - \exp\left(-\int_0^t \lambda(u)du\right), \tag{5.4}$$

mimicking formula (3.10) for the survival function. Solving (5.4) for λ, we find that (5.4) holds if we define

$$\lambda(t) \cdot dt := \frac{P(T \in dt, X_T = 1)}{1 - P(T \leq t, X_T = 1)} \tag{5.5}$$

It also follows from representations (3.11) and (5.4) that

$$\alpha_{01}(t) = \left(1 + \frac{P(T \leq t, X_T = 2)}{P(T > t)}\right) \cdot \lambda(t).$$

Later, in Section 5.3, we derive the subdistribution hazards framework by suitably *stopping* the original competing risks process, which will also be useful for handling time-dependent covariates. The interpretation of the last display is that the subdistribution hazard for event type 1 is weighted down as compared to the cause-specific hazard $\alpha_{01}(t)$ with a weighting that is time-dependent and also depends on the competing events. Note that the subdistribution hazard is weighted *down*, because (5.4) implies that

$$P(X_T = 1) = 1 - \lim_{t \to \infty} \exp\left(-\int_0^t \lambda(u)du\right) = 1 - P(X_T = 2), \tag{5.6}$$

whereas the limit of exp of the negative of a usual cumulative all-cause hazard will be zero, which is the limit of a survival function as time goes to infinity.

We discuss proportional subdistribution hazards modelling (Fine and Gray, 1999) in Section 5.3,

$$\lambda_i(t; Z_i) = \lambda_0(t) \cdot \exp\left(\gamma \cdot Z_i\right), \ i = 1, \ldots, n, \tag{5.7}$$

where Z_i is as in (5.1), γ is a $1 \times p$ vector of regression coefficients, and $\lambda_0(t)$ is an unspecified, non-negative baseline *subdistribution* hazard function. The results have a direct probability interpretation in terms of the cumulative incidence function. Note that in general $\gamma \neq \beta_{0j}$, $j = 1, 2$.

Both modelling cause-specific hazards and modelling subdistribution hazards have their merits. We find that only the subdistribution hazard analysis allows for a *direct* probability interpretation. The analyses of the cause-specific hazards also allow for a probability interpretation, but the interpretation requires greater care. It is, however, only through analysing all of the cause-specific hazards that we understand *why* we see a certain effect on the probability functions. We already discussed this tradeoff in the first analysis of the pneumonia data in Section 4.3.

The data analyses of Sections 5.2.2 and 5.3.3 illustrate that proportional cause-specific hazards modelling and proportional subdistribution hazards modelling address different aspects of the data, the latter analysis offering a synthesis of the effects on the different cause-specific hazards. However, model (5.7) is misspecified if models (5.1) hold. Moreover, the all-cause hazard $\alpha_{0\cdot}(t) = \alpha_{01}(t) + \alpha_{02}(t)$ will in general not follow a Cox model if the cause-specific hazards do. In Section 5.4, we discuss that a misspecified model still provides a consistent estimate, although not of the regression coefficient of the misspecified model, but of the so-called least false parameter (Hjort, 1992), a time-averaged hazard ratio. For a concrete data analysis, this entails that a proportional effect of a covariate on the cause-specific hazards, on the subdistribution hazard, or on the all-cause hazard is not claimed. But one would profit from the simple structure of a proportional hazards model because its results display an average effect on the respective hazard scale. This 'agnostic' point of view (Hjort, 1992) towards model assumptions is illustrated in the data analyses of Sections 5.2.2 and 5.3.3.

Goodness-of-fit methods and regression models that go beyond the familiar proportional hazards assumption are briefly discussed in Sections 5.5 and 5.6.

5.2 Proportional cause-specific hazards models

We consider the proportional cause-specific hazards models (5.3) with cause-specific covariate vectors $Z_{0j;i}$. We introduce estimation and prediction in Section 5.2.1. It is important to recognize that we may compute survival probabilities and cumulative incidence functions of the observable competing risks process under model assumption (5.3), that is, if we model all cause-specific hazards. In practice, this requires the analysis of both cause-specific hazards. This is often overlooked in concrete data analyses. We apply these methods in Section 5.2.2 to both simulated and real data. We later also analyse the real data, which served as a template for the simulated data, in the Exercises of Section 5.7.

5.2.1 Estimation

We first consider estimation of the regression coefficients. Next, we consider model-based estimation of the cumulative cause-specific baseline hazards, and finally prediction of the cumulative incidence functions is introduced.

Estimation of the regression coefficients

Recall from Section 4.1 that we have individual cause-specific counting processes $N_{0j;i}(t)$ that count whether we have observed a type j event for individual i during the time interval $[0, t]$. Also recall that the individual at-risk process is $Y_{0;i}(t)$. The respective processes aggregated over all individuals i, $i = 1, \ldots, n$ have been defined as $N_{0j}(t)$ and $Y_0(t)$, respectively. Finally, $N_{0.}(t) = N_{01}(t) + N_{02}(t)$ is the number of observed transitions out of the initial state 0 during the time interval $[0, t]$.

We define the weighted risk set

$$S_{0j}^{(0)}(\beta, t) := \sum_{i=1}^{n} \exp\left(\beta \cdot Z_{0j;i}\right) \cdot Y_{0;i}(t). \tag{5.8}$$

We explain the meaning of this weighted risk set in a moment. Note that the (standard) notation $S_{0j}^{(0)}(\beta, t)$ should not be confused with a survival function $P(T > t)$, which is sometimes denoted as $S(t)$. Also note that $S_{0j}^{(0)}$ really depends on the transition type $0 \to j$ through the cause-specific covariates $Z_{0j;i}$.

Estimation of β is based on the partial likelihood

$$L(\beta) = \prod_{t} \prod_{i=1}^{n} \prod_{j=1}^{2} \left(\frac{\exp\left(\beta \cdot Z_{0j;i}\right)}{S_{0j}^{(0)}(\beta, t)} \right)^{\Delta N_{0j;i}(t)}, \tag{5.9}$$

where the first product is over all times t, where an event of type 1 or 2 was observed for some individual. The interpretation and heuristic derivation of (5.9) is that $\exp\left(\beta \cdot Z_{0j;i}\right)/S_{0j}^{(0)}(\beta, t)$ is the probability that it is exactly individual i who fails from event type j given that an event of type j is observed at time t: it follows from the definition of the cause-specific hazards in (3.5) that the probability that we observe a type j event in the very small time interval dt, given that we know all prior events of either type and censoring events, is

$$P(dN_{0j}(t) = 1 \mid \text{Past}) = \left(\sum_{i=1}^{n} Y_{0;i}(t) \cdot \alpha_{0j;i}(t; Z_i) \right) \cdot dt, \tag{5.10}$$

assuming that no two individuals experience an event at the same time. In (5.10), we have briefly written 'Past' for prior events of either type and censoring events. Using model assumption (5.3) and Definition (5.8), quantity (5.10) equals $\alpha_{0j;0}(t) \cdot S_{0j}^{(0)}(\beta, t) \cdot dt$. Analogously, the respective conditional probability that we observe a type j event in the very small time interval dt precisely for individual i is $Y_{0;i}(t)\alpha_{0j;0}(t) \cdot \exp\left(\beta \cdot Z_{0j;i}\right) dt$. Hence,

P(Individual i observed to fail from cause j at $t \mid$

$$\Delta N_{0j}(t) = 1, \text{Past}) = \frac{Y_{0;i}(t) \cdot \exp\left(\beta \cdot Z_{0j;i}\right)}{S_{0j}^{(0)}(\beta, t)}. \tag{5.11}$$

Note that we used a related argument on the ratio of an individual hazard and an appropriate sum of hazards in step 3 in the competing risks simulation algorithm of Section 3.2. The leading factor $Y_{0;i}(t)$ in the numerator is dropped from the partial likelihood $L(\beta)$ in (5.9) as an exponent $\Delta N_{0j;i}(t) \neq 0$ implies $Y_{0;i}(t) = 1$. $L(\beta)$ is a *partial* likelihood, because the baseline cause-specific hazards $\alpha_{0j;0}(t)$ have canceled out in (5.11) and in (5.9). In his seminal paper, Cox (1972) suggested that statistical inference for β could be based on maximizing $L(\beta)$, because the baseline hazard functions have been left completely unspecified and, hence, the time intervals between observed event times should not contain any information on β.

The question of what type of likelihood is displayed by $L(\beta)$ has been considered in detail in the literature following Cox' paper. In fact, martingale theory shows that $L(\beta)$ enjoys large sample theory that shares a lot of the flavor of standard maximum likelihood theory, and that we may use for (approximate) inference in practice. Gill (1984) gives a very accessible account of this issue.

We consider large sample properties of $\widehat{\beta}$. Let us write $\beta^{(0)}$ for the true parameter vector in (5.3). Then $\sqrt{n}(\widehat{\beta} - \beta^{(0)})$ approaches a multinormal distribution with mean zero (Andersen and Gill, 1982; Andersen and Borgan, 1985). In order to write down an estimator of the covariance matrix, we need to introduce some additional notation. Readers content with the fact that the covariance matrix can be estimated may very well skip this part. We write

$$S_{0j}^{(1)}(\beta, t) := \sum_{i=1}^{n} Z_{0j;i} \cdot \exp\left(\beta \cdot Z_{0j;i}\right) \cdot Y_{0;i}(t), \, j = 1, 2, \qquad (5.12)$$

and

$$\mathbf{E}_{0j}(\beta, t) = \frac{S_{0j}^{(1)}(\beta, t)}{S_{0j}^{(0)}(\beta, t)}. \qquad (5.13)$$

If β is a $1 \times q$ vector and $Z_{0j;i}$ are $q \times 1$ vectors, then both $S_{0j}^{(1)}(\beta, t)$ and $\mathbf{E}_{0j}(\beta, t)$ are $q \times 1$ vectors. We also write

$$S_{0j}^{(2)}(\beta, t) := \sum_{i=1}^{n} Z_{0j;i} Z_{0j;i}^{\top} \cdot \exp\left(\beta \cdot Z_{0j;i}\right) \cdot Y_{0;i}(t), \, j = 1, 2, \qquad (5.14)$$

a $q \times q$ matrix, and

$$\mathbf{V}_{0j}(\beta, t) = \frac{S_{0j}^{(2)}(\beta, t)}{S_{0j}^{(0)}(\beta, t)} - \mathbf{E}_{0j}(\beta, t)\mathbf{E}_{0j}(\beta, t)^{\top}, \, j = 1, 2, \qquad (5.15)$$

a $q \times q$ matrix, too. It can then be observed that the matrix of the second-order partial derivatives of the log partial likelihood $\ln L(\beta)$ is the negative of

$$\mathcal{J}_{\tau}(\beta) = \sum_{j=1}^{2} \sum_{T_i \wedge C_i \leq \tau} \mathbf{V}_{0j}(\beta, T_i \wedge C_i) \Delta N_{0j}(T_i \wedge C_i), \qquad (5.16)$$

where τ is taken to be the largest observed event time. Also recall that $T_i \wedge C_i$ is the minimum of individual i's event time T_i and censoring time C_i. We estimate the covariance matrix of $(\widehat{\beta} - \beta^{(0)})$ by $(\mathcal{J}_{\tau}(\widehat{\beta}))^{-1}$. As illustrated in Section 5.2.2, the result is used to compute variance estimates, confidence intervals, and p-values for the regression coefficients. It also follows that the asymptotic distribution of the Wald test statistic

$$(\widehat{\beta} - \beta^{(0)}) \mathcal{J}_\tau(\widehat{\beta})(\widehat{\beta} - \beta^{(0)})^\top$$

for the hypothesis $\beta = \beta^{(0)}$ is chi-square with q degrees of freedom. In the examples below, we show that standard R software also reports a likelihood ratio test and a score test, which have the same asymptotic distribution. The score test statistic is the usual log-rank test statistic in the case of standard single endpoint survival data, a single categorical covariate and $\beta^{(0)} = 0$. A discussion of the relative merits of these tests can be found in Therneau and Grambsch (2000, Section 3.4) and Hosmer et al. (2008, Section 3.3), among others. Hosmer et al. summarize that these tests are usually quite similar; if they disagree, conclusions should be based on the likelihood ratio test.

It is worthwhile to note that the partial likelihood $L(\beta)$ splits into two factors $L(\beta_{01})$ and $L(\beta_{02})$,

$$L(\beta_{0j}) = \prod_t \prod_{i=1}^n \left(\frac{\exp(\beta_{0j} \cdot Z_i)}{\sum_{l=1}^n \exp(\beta_{0j} \cdot Z_l) \cdot Y_{0;l}(t)} \right)^{\Delta N_{0j;i}(t)}, \quad j = 1, 2, \quad (5.17)$$

depending on the respective β_{0j} only if we assume two completely separate models for the cause-specific hazards; that is, $\beta = (\beta_{01}, \beta_{02})$ with no common effect on these hazards. Obviously, the maximum likelihood estimate may be obtained by separately maximizing $\widehat{\beta}$ by separately maximizing $L(\beta_{01})$ and $L(\beta_{02})$. As a further consequence, $q = p + p$, and the $p \times p$ blocks in $(\mathcal{J}_\tau(\widehat{\beta}))^{-1}$ corresponding to $\widehat{\beta}_{0j}$ are those that one would get from analysing $L(\beta_{0j})$ only. We illustrate this in Section 5.2.2.

Breslow estimator of the cumulative cause-specific hazards and prediction

Let $\widehat{\beta}$ be the estimator that results from maximizing $L(\beta)$. A Nelson-Aalen-type estimator of the cumulative cause-specific baseline hazards is

$$\widehat{A}_{0j;0}(t) := \sum_{T_i \wedge C_i \leq t} \frac{\Delta N_{0j}(T_i \wedge C_i)}{S_{0j}^{(0)}(\widehat{\beta}, T_i \wedge C_i)}, \quad j = 1, 2. \quad (5.18)$$

In order to motivate this estimator, note that, following (5.10), we have seen that $P(dN_{0j}(t) = 1 \,|\, \text{Past}) = \alpha_{0j;0}(t) \cdot S_{0j}^{(0)}(\beta, t) \cdot dt$. In the absence of covariates, that is, for a homogeneous sample, this quantity becomes $\alpha_{0j}(t) \cdot Y_0(t)$, which motivates the increments $\Delta N_{0j}(t)/Y_0(t)$ of the Nelson-Aalen estimator (4.8) (cf. our discussion preceding (4.8)). Under model assumption (5.3), we substitute Y_0 by $S_{0j}^{(0)}$ and use the estimator $\widehat{\beta}$ for the unknown regression coefficients which results in (5.18).

The Nelson-Aalen-type estimator (5.18) is often called Breslow estimator (Breslow, 1972). In proportional hazards analyses of standard (single end-point) survival data, the all-cause baseline hazard is frequently not considered as it cancels out from the partial likelihood. This is convenient in that the data analyst does not need to consider the baseline hazard in order to study the impact of the covariates on survival. However, this may also be seen to be unfavorable as it tempts one to neglect the baseline hazard, which determines the survival function together with $\beta \cdot Z_i$. If anything, looking at the cause-specific baseline hazards is even more important in a competing risks analysis as cause-specific regression coefficients of a similar magnitude may have quite different effects on the cumulative incidence functions, depending on the magnitude of the cause-specific baseline hazards. We illustrate this in the examples below, in particular in our discussion of the model specifications for the data simulation in Section 5.2.2.

We now consider prediction under model (5.3) for some (future) individual $\tilde{\imath}$ with individual covariate $Z_{\tilde{\imath}} = z$, which we may rewrite as individual cause-specific covariates $Z_{0j;\tilde{\imath}} = z_{0j}$. In the following, we drop the index $\tilde{\imath}$ for notational convenience. The individual cumulative cause-specific hazards may be predicted as

$$\widehat{A}_{0j}(t; z) = \widehat{A}_{0j;0}(t) \cdot \exp\left(\widehat{\beta} \cdot z_{0j}\right), \, j = 1, 2, \tag{5.19}$$

and the individual cumulative all-cause hazard may be predicted as

$$\widehat{A}_{0\cdot}(t; z) = \widehat{A}_{01}(t; z) + \widehat{A}_{02}(t; z). \tag{5.20}$$

To arrive at a predictor of the individual all-cause survival function, we substitute the increments of the usual Nelson-Aalen estimator in Equation (4.13) by the increments $\Delta\widehat{A}_{0\cdot}(t; z)$:

$$\widehat{P}(T > t \,|\, z) = \prod_{T_i \wedge C_i \leq t} \left(1 - \Delta\widehat{A}_{0\cdot}(T_i \wedge C_i; z)\right) \tag{5.21}$$

with increments

$$\Delta\widehat{A}_{0\cdot}(t; z) = \sum_{j=1}^{2} \frac{\Delta N_{0j}(T_i \wedge C_i)}{S_{0j}^{(0)}(\widehat{\beta}, T_i \wedge C_i)} \cdot \exp\left(\widehat{\beta} \cdot z_{0j}\right). \tag{5.22}$$

Recalling the estimator of the cumulative incidence function in Equation (4.18), we arrive at an individual predictor of this probability by substituting the Kaplan-Meier estimator by (5.21) and the increment of the cause-specific Nelson-Aalen estimator by $\Delta\widehat{A}_{0j}(t; z)$:

$$\widehat{P}(T \leq t, X_T = j \,|\, z) =$$

$$\sum_{T_i \wedge C_i \leq t} \widehat{P}(T > (T_i \wedge C_i) - \,|\, z) \cdot \frac{\Delta N_{0j}(T_i \wedge C_i)}{S_{0j}^{(0)}(\widehat{\beta}, T_i \wedge C_i)} \cdot \exp\left(\widehat{\beta} \cdot z_{0j}\right), \tag{5.23}$$

for $j = 1, 2$.

Large sample properties such as those discussed above for $\widehat{\beta}$ are also available for the cumulative baseline estimator (5.18) and the predicted quantities (5.19)–(5.23). However, the formulae become increasingly involved. We have therefore chosen to refer the interested reader to Andersen et al. (1993, Sections VII.2.2—VII.2.3) for a general treatment and to Rosthøj et al. (2004) for the present case of competing risks. Implementation in R has been made available during the writing of this book via the `mstate` package (de Wreede et al., 2010, 2011) and is illustrated below. Alternatively, in situations where variance formulae become increasingly complex, one may consider *bootstrapping* the data, which approximates the asymptotic distribution by a simulation experiment; see Appendix A. We also note that Cheng et al. (1998) considered a related resampling scheme, when the aim is prediction of the cumulative incidence function.

5.2.2 Examples

We consider both simulated data and real data. To begin, we investigate in some detail the model specification for the simulated data, which is motivated by a study on infectious complications in stem-cell transplanted patients. We find that even with perfect knowledge of the data-generating cause-specific hazards care must be displayed when drawing conclusions on the cumulative event probabilities. The real data that serve as a template are later analysed in the Exercises of Section 5.7.

Next, we analyse the simulated data. Our main workhorse is the `coxph`-function from the `survival` package. We provide a guided tour of the typical steps often taken in a competing risks analysis. Besides a standard first-event analysis, which does not distinguish between the competing event states, two ways of analysing cause-specific hazards in practice are discussed: one approach, which often suffices in practice, fits two separate Cox models. The other approach requires data duplication and coding of cause-specific covariates as in (5.3).

We find that the model specification can be recovered from the simulated data, i.e., the regression coefficients of the proportional cause-specific hazards models *and* the cumulative baseline hazards. As we illustrate below, having these two types of information is crucial for prediction.

Finally, we revisit the real data analyses of Sections 4.3–4.4.

Simulated data

Model specification, motivation, and analysis of the model specification

Simulating competing risks data from model (5.3) runs along the lines of the simulation algorithm of Section 3.2. The only difference is that we first need to determine an individual's baseline covariate values Z_i. This is not specific to competing risks, but a step one also has to take in simulating standard single endpoint survival data; see, e.g., Bender et al. (2005); Burton et al. (2006). We illustrate this with a simple binary covariate $Z_i \in \{0,1\}$. We find that interpretation of the analyses is already challenging in this simple example. The example is also important in that it is often a (binary) treatment or exposure information that is being investigated.

Consider the following specification of the cause-specific baseline hazards:

$$\alpha_{01;0}(t) = \alpha_{01}(t; Z_i = 0) = \frac{0.09}{t+1}, \tag{5.24}$$

$$\alpha_{02;0}(t) = \alpha_{02}(t; Z_i = 0) = 0.024 \cdot t, \, i = 1, \ldots, n. \tag{5.25}$$

We assume that $Z_i = 1$ displays separate reducing effects on the α_{0j}s:

$$\alpha_{01}(t; Z_i = 1) = 0.825 \cdot \alpha_{01;0}(t), \tag{5.26}$$

$$\alpha_{02}(t; Z_i = 1) = 0.2 \cdot \alpha_{02;0}(t), \, i = 1, \ldots, n, \tag{5.27}$$

with cause-specific regression coefficients $\beta_{01} = \ln 0.825 \approx -0.19$ and $\beta_{02} = \ln 0.2 \approx -1.6$. As explained following (5.3), we may reformulate these models with one vector $\beta = (\ln 0.825, \ln 0.2)$ and cause-specific covariates $Z_{01;i} = (Z_i, 0)^\top$ and $Z_{02;i} = (0, Z_i)^\top$:

$$\alpha_{01}(t; Z_i = 1) = \frac{0.09}{t+1} \cdot \exp\left((\ln 0.825, \ln 0.2) \cdot Z_{01;i}\right) \tag{5.28}$$

$$\alpha_{02}(t; Z_i = 1) = 0.024 \cdot t \cdot \exp\left((\ln 0.825, \ln 0.2) \cdot Z_{02;i}\right). \tag{5.29}$$

How to *code* cause-specific covariate vectors is illustrated in the analysis of the data frame x1 below. Before we simulate and analyse data from these models, we investigate their specification in more detail. The simulation experiment that we analysed in Section 4.2 had the air of a 'toy example'. In contrast, the current model, which is taken from Beyersmann et al. (2009), is motivated by the prospective cohort study ONKO-KISS (Meyer et al., 2007). ONKO-KISS assesses risk factors for the occurrence of bloodstream infections (BSI) during neutropenia, a condition where patients have a low count of white blood cells; these are the cells that primarily avert infections. Patients treated for severe hematological diseases by peripheral blood stem-cell transplantation become neutropenic immediately after the transplantation. Occurrence of BSI during neutropenia constitutes a severe complication and substantially endangers the success of the therapy. Allogeneic transplant type is considered to be a risk factor for the occurrence of BSI as opposed to autologous transplants.

This situation is one of competing risks: every patient enters state 0 of the competing risks model in Figure 3.1 at time origin $t = 0$ following transplantation. The event state 1 of interest is occurrence of BSI during neutropenia; observation of BSI may be precluded by occurrence of the competing event state 2 (i.e., end of neutropenia without prior BSI). Thus, T denotes the time until BSI or end of neutropenia, whichever comes first, and $X_T = 1$ denotes the failure type of interest (i.e., BSI), and $X_T = 2$ denotes the competing failure type (i.e., end of neutropenia). Here, $X_T = 2$ is a combined competing endpoint, because neutropenia may be ended either alive or dead. Not distinguishing the vital status looks awkward, but can be justified in this concrete setting by the fact that hardly a patient died in neutropenia without prior BSI.

Our model specifications have been chosen to roughly mimic the effect of allogeneic transplant type. Let $Z_i = 1$, if patient i had received an allogeneic transplant, and $Z_i = 0$ for an autologous transplant. In the analysis of the data (Beyersmann et al., 2007), it was found that allogeneic transplants displayed a reducing effect on both cause-specific hazards, which is reflected in Equations (5.26)–(5.29). It was also found that the proportion of BSI patients was higher in the allogeneic group than in the autologous group. The latter result is in line with the assessment of allogeneic transplants as a risk factor. The reducing effect of allogeneic transplant on the cause-specific infection hazard in (5.26) appears to contradict this. This is a typical competing risks phenomenon, which is well understood by looking at plots of the model quantities in connection with the simulation algorithm of Section 3.2:

Figure 5.1 displays the cause-specific hazards. We note three things: first,

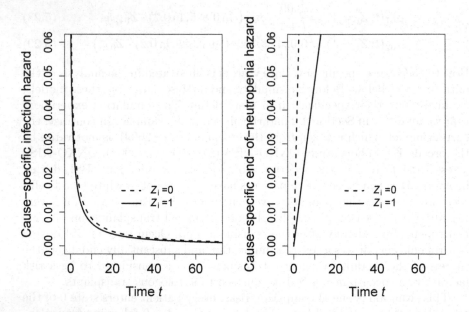

Fig. 5.1. *Model specification for simulation.* Left plot: Cause-specific hazards $\alpha_{01}(t; Z_i)$ for the event of interest. Right plot: Competing cause-specific hazards $\alpha_{02}(t; Z_i)$.

$Z_i = 1$ reduces both cause-specific hazards and, as a consequence of that, the all-cause hazard $\alpha_0.(t)$. Hence, individuals with $Z_i = 1$ stay longer in the initial state of the competing risks processes; this is relevant for step 2 of the simulation algorithm, in which an individual's event time is determined. Second, this reducing effect is stronger on the competing cause-specific hazard in the sense that $\beta_{02} < \beta_{01}$; this is relevant for step 3 of the simulation algorithm in which an individual's event type is determined, inasmuch as (assuming all hazards to be positive)

$$\beta_{02} < \beta_{01}$$
$$\Longleftrightarrow e^{\beta_{01}}\alpha_{01;0}(t) + e^{\beta_{02}}\alpha_{02;0}(t) < e^{\beta_{01}}\alpha_{01;0}(t) + e^{\beta_{01}}\alpha_{02;0}(t)$$
$$\Longleftrightarrow \frac{\alpha_{01;0}(t)}{\alpha_{01;0}(t) + \alpha_{02;0}(t)} < \frac{e^{\beta_{01}}\alpha_{01;0}(t)}{e^{\beta_{01}}\alpha_{01;0}(t) + e^{\beta_{02}}\alpha_{02;0}(t)}$$
$$\Longleftrightarrow P(X_{T_i} = 1 \mid T_i = t, Z_i = 0) < P(X_{T_i} = 1 \mid T_i = t, Z_i = 1). \quad (5.30)$$

This means that the binomial event type 1 probability is smaller in the baseline group at any time t. Third, judged from the magnitudes displayed in Figure 5.1 and except for early time points, the competing cause-specific hazard is the major hazard regardless of the covariate value. We should therefore expect fewer type 1 events than type 2 events in both groups.

Consequences for step 2 and step 3 are displayed in Figure 5.2. As stated above, individuals with $Z_i = 1$ stay longer in the initial state. At any time that such an individual leaves the initial state, the individual's binomial event type 1 probability is greater than in the baseline group. As a consequence, the cumulative incidence functions will increase later for individuals with $Z_i = 1$ in the current set-up; in this group, we *eventually* see more type 1 events. This effect is illustrated in Figures 5.3 and 5.4.

Going back to the motivation behind the current model specification, the preceding analysis of these specifications entails that allogeneic transplant type is a risk factor for infection, because its reducing effect on the cause-specific infection hazard is less pronounced than its reducing effect on the competing cause-specific hazard. This also entails that we may initially see both fewer events of either type and fewer infection events in the allogeneic group. However, every individual will eventually experience a first event, and the proportion of infection events will eventually be higher in the allogeneic group.

This kind of reasoning is what is often needed in the interpretation of results from competing risks analyses; cf. our analyses of the simulated data below and also Section 4.3. However, a cautionary note is in place: so far, our discussion of the different magnitudes of the cause-specific hazards for event 1 and event 2 has been mostly restricted to noting that most individuals will experience the competing event 2 in either group. It should also be noted that the major effect is displayed by β_{02}, which acts on the (eventually) major hazard. Therefore, it is probably not surprising that $P(T_i \leq t, X_{T_i} = 2 \mid Z_i = 1)$ runs below $P(T_i \leq t, X_{T_i} = 2 \mid Z_i = 0)$. But this kind of reasoning may

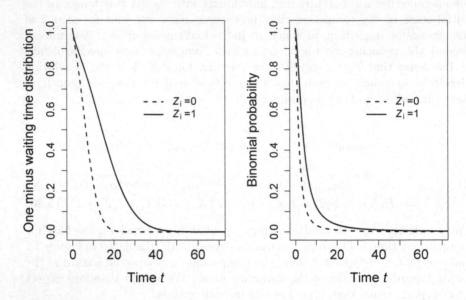

Fig. 5.2. *Model specification for simulation.* Left plot: Probability $P(T > t \mid Z_i)$ of staying in the initial state of the competing risks process. Right plot: Binomial event type 1 probability $P(X_{T_i} = 1 \mid T_i = t, Z_i = \cdot)$.

not suffice: note from Figure 5.2 (right plot) that $P(X_{T_i} = 2 \mid T_i = t, Z_i = \cdot)$ equals one minus the displayed curves. Hence, this probability is eventually much larger than $P(X_{T_i} = 1 \mid T_i = t, Z_i = \cdot)$. This is because $\alpha_{02}(t; Z_i = \cdot)$ is eventually the major hazard, cf. (5.30) and step 3 of the simulation algorithm of Section 3.2. It follows that there may be situations with regression coefficients $\beta_{02} < \beta_{01} < 0$ and reversed constellations of the *plateaus* of the cumulative incidence functions. The left plot in Figure 5.5 displays such a situation, where we have kept model specifications (5.24)–(5.27) except for changing β_{02} from $\ln 0.2$ to $\ln 0.7$. The right plot in Figure 5.5 displays the situation where $\beta_{01} = \beta_{02} = \ln 0.825$. Common to both model changes is that the effect on $\alpha_{02}(t)$ is less pronounced. We discuss the situation of a common reducing effect on both cause-specific hazards first.

In the right plot of Figure 5.5, $Z_i = 1$ displays a common reducing effect on both cause-specific hazards. As a consequence, and as described before, individuals stay in the initial state of the competing risks process longer. As $\beta_{01} = \beta_{02}$, the binomial probabilities $P(X_{T_i} = j \mid T_i = t, Z_i = \cdot)$, $j = 1, 2$, remain unchanged. It is now important to note that eventually $P(X_{T_i} = 1 \mid T_i = t, Z_i = \cdot) < P(X_{T_i} = 2 \mid T_i = t, Z_i = \cdot)$ and that $P(X_{T_i} = 2 \mid T_i = t, Z_i = \cdot)$ increases with t (cf. Figure 5.3). Both facts are an immediate consequence of $\alpha_{02}(t; Z_i = \cdot)$ being the major hazards.

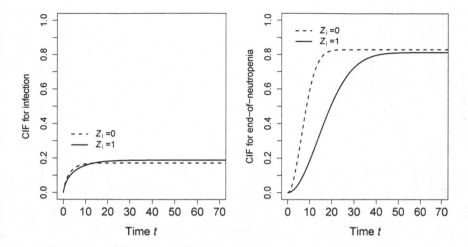

Fig. 5.3. *Model specification for simulation.* Cumulative incidence functions (CIF) for $X_T = 1$ (left plot) and for $X_T = 2$ (right plot). The left plot is redisplayed in Figure 5.4 with different scales for the axes.

An event in the group with $Z_i = 1$ tends to happen at later times, therefore it will more likely be of the competing type 2 as compared to the baseline group. Therefore, we eventually see $P(T_i \leq t, X_{T_i} = 2 \,|\, Z_i = 1)$ to run above $P(T_i \leq t, X_{T_i} = 2 \,|\, Z_i = 0)$ in Figure 5.5 (right plot).

The left plot of Figure 5.5 displays a situation with similar constellation of the plateaus of the cumulative incidence functions, but we have again $\beta_{02} < \beta_{01}$ as in our original model specifications. However, unlike in the original set-up, $\beta_{02} = \ln 0.7$ is now close to $\beta_{01} = \ln 0.825$, and hence this situation is closer to the one of a common effect. Still, as $\beta_{02} < \beta_{01}$, we have for the binomial type 1 probabilities that

$$P(X_{T_i} = 1 \,|\, T_i = t, Z_i = 0) < P(X_{T_i} = 1 \,|\, T_i = t, Z_i = 1),$$

cf. (5.30). The difference, however, is now less pronounced, because $\beta_{02} = \ln 0.7$ is now closer to zero. Individuals with $Z_i = 1$ stay in the initial state slightly longer. Although they are exposed to a slightly increased binomial type 1 probability, the major effect seen is that such an individual is exposed to a higher binomial type 2 probability at time T_i as compared to an earlier point in time.

The bottom line is that the previous situation resembles more the common effect situation. In an actual data analysis, it might be impossible to distinguish between them. In fact, the cause-specific effect on $\alpha_{01}(t)$ was only just significant in the analysis of the ONKO-KISS data that motivated specifications (5.24)–(5.27), the right end of the 95% confidence interval for the cause-

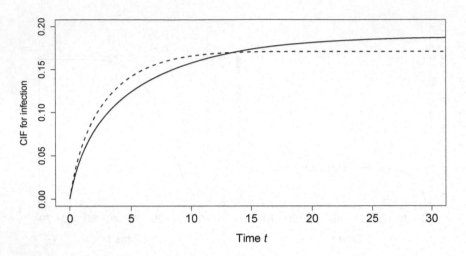

Fig. 5.4. *Model specification for simulation.* Cumulative incidence function (CIF) for $X_T = 1$ as in Figure 5.3, but with different scales for the axes. See Figure 5.9 for an empirical analogue of the present figure.

specific hazard ratio being 0.99. However, as always, there may be 'borderline cases' even in practice. Readers are encouraged to both analyse cause-specific *baseline* hazards in a real data analysis and to experiment with cause-specific hazards specifications for simulation purposes.

Simulation and analysis of simulated data

Roughly mimicking the ONKO-KISS study, we simulate 1500 individuals; for reasons of simplicity, we assume 750 individuals to have covariate value 0, also roughly mimicking ONKO-KISS. The simulation follows the algorithm explained in detail in Section 3.2. In step 1 of the algorithm, individuals with covariate value 0 have cause-specific hazards specifications (5.24)–(5.25); individuals with covariate value 1 have cause-specific hazards specifications (5.26)–(5.27). In step 2 of the algorithm, we use numerical inversion as explained towards the end of Section 3.2. Finally, still mimicking ONKO-KISS, we simulate light censoring, uniformly distributed on $(0, 100)$, which is independent from everything else. We also define an id-variable $1:1500$. We aggregate this data in a data frame x. The entry for individual 1 is displayed below:

```
> x[1, ]
```

```
    id        T X.T Z       C   TandC status
1    1 8.688352   2 0 89.5042 8.688352      2
```

Entry id is the id-variable, entry T the event time, X.T the competing risks failure type X_T (either 1 or 2), Z contains the covariate value, and C is the

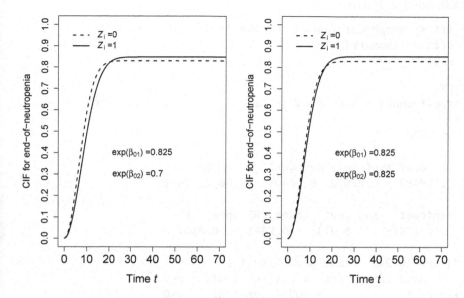

Fig. 5.5. *Model specification for simulation.* Cumulative incidence functions (CIF) for $X_T = 2$ with baseline hazards as in (5.24) and (5.25), cause-specific regression coefficient β_{01} as in (5.26), but different β_{02}.

censoring time. The remaining two entries are derived from this information. Entry `TandC` is the censored failure time $T \wedge C$, and `status` is the observed failure status $\mathbf{1}(T \leq C) \cdot X_T$,

```
> x$TandC <- pmin(x$T,x$C)
> x$status <- as.numeric(x$T<=x$C)*x$X.T
```

Censoring is light:

```
> sum(x$T > x$C) / length(x$T)
```

```
[1] 0.1066667
```

We first run a first-event analysis, not distinguishing between the two competing event types. Next, we analyse the cause-specific hazards by fitting two separate Cox models, which is tantamount to assuming that there is no common effect of Z on the α_{0j}s. We then analyse a duplicated data set that allows for directly maximizing the partial likelihood $L(\beta)$ of (5.9). Finally, we compute the Breslow estimators as well as the nonparametric Nelson-Aalen and Aalen-Johansen estimators of Chapter 4; in this last step, the importance of the cause-specific baseline hazards for interpreting results from cause-specific Cox models is again pointed out. We also predict the cumulative incidence functions based on Cox models for the cause-specific hazards.

First-event analysis We use the `survival` package (Therneau and Grambsch, 2000; Lumley, 2004):

```
> fit <- coxph(Surv(TandC, status != 0) ~ Z, data = x)
> sfit <- summary(fit)
> sfit

Call:
coxph(formula = Surv(TandC, status != 0) ~ Z, data = x)

  n= 1500

      coef exp(coef) se(coef)      z Pr(>|z|)
Z -1.28482   0.27670  0.06405 -20.06   <2e-16


  exp(coef) exp(-coef) lower .95 upper .95
Z    0.2767      3.614    0.2441    0.3137

Rsquare= 0.242    (max possible= 1 )
Likelihood ratio test= 415.1  on 1 df,    p=0
Wald test              = 402.4  on 1 df,    p=0
Score (logrank) test = 442.1  on 1 df,    p=0
```

The `Surv`-function has been explained in the analysis of a survival function in Section 4.2. The argument `status!=0` specifies that we consider any observed competing event (observed at time `TandC`) as an event. The function `coxph` fits a proportional hazards model to these data; covariates are on the right of the `~`-operator, and `data=x` requests that the variables named in the formula `Surv(TandC,status!=0)~Z` are interpreted in the data frame x.

This model is misspecified, as Z displays different effects on the cause-specific hazards. As a consequence, the effect of Z on the all-cause hazard does not follow a proportional hazards model. Nevertheless, we can interpret the analysis in a meaningful way. We first look at the different results returned by `summary(fit)`: The all-cause hazard ratio $\alpha_{0\cdot;i}(t; Z_i = 1)/\alpha_{0\cdot;i}(t; Z_i = 0)$ is estimated as 0.277, $\ln 0.277 \approx -1.285$. The 95% confidence interval for the all-cause hazard ratio is estimated to be [0.244, 0.314]. This interval is based on the asymptotic normality of the maximum likelihood estimate: `fit$coefficients` contains the maximum likelihood estimate, `fit$var` the variance estimate. We have

```
> exp(fit$coefficients + qnorm(0.975) *sqrt(fit$var))

          [,1]
[1,] 0.3137099

> exp(fit$coefficients - qnorm(0.975) *sqrt(fit$var))
```

```
         [,1]
[1,] 0.2440563
```

The z-statistic reported in the summary of `fit` displayed above is `fit$coefficients` divided by `sqrt(fit$var)`. The value reported by `Rsquare` is an attempt to measure the predictive 'value' of the model. Assessing prediction for both standard survival data (Gerds et al., 2008) and competing risks (Saha and Heagerty, 2010; Schoop et al., 2011) is still an active research field, and there appears to be hardly any work being undertaken for more complex multistate models. Therefore, we do not further discuss `Rsquare` in this book. The tests reported at the end of the output of `summary(fit)` are as discussed towards the end of Section 5.2.1 for the hypothesis $\beta = 0$.

Let us ignore for a moment that the model is misspecified. In fact, we would first need to detect the misspecification in an analysis of non-simulated data (cf. Section 5.5). The above analysis finds that the waiting time T in the initial state is reduced by covariate value 1. This finding is correct, as covariate value 1 reduces both cause-specific hazards; it therefore also reduces the all-cause hazard. We also find that the estimated all-cause hazard ratio 0.277 is close to the cause-specific hazard ratio 0.2 for the competing event. This is meaningful because $\alpha_{02}(t)$ is the major hazard. With knowledge of the model specifications (5.26)–(5.27), it is also not unexpected that the estimated all-cause hazard ratio is somewhat larger than 0.2, as the reducing effect on $\alpha_{01;0}(t)$ is less pronounced. In Section 5.4, we describe in more detail that such an interpretation of results from a misspecified model actually discusses estimates of a *time-averaged* hazard ratio.

Analysis of the cause-specific hazards by fitting two separate Cox models The present analysis assumes that there is no common effect of Z on the α_{0j}s. The analysis of $\alpha_{01}(t)$ only treats type 1 events as events and handles competing type 2 events and the usual censorings alike. The analysis of $\alpha_{02}(t)$ does it the other way round. This is in complete analogy to the Nelson-Aalen estimates of the cumulative cause-specific hazards (cf. (4.1)). This is the analysis of the cause-specific hazard for the event of interest:

```
> summary(coxph(Surv(TandC, status == 1) ~ Z, data = x))

  n= 1500

     coef exp(coef) se(coef)      z Pr(>|z|)
Z -0.1682    0.8451   0.1274  -1.32    0.187

  exp(coef) exp(-coef) lower .95 upper .95
Z    0.8451      1.183    0.6583     1.085
```

The estimate 0.845 of the cause-specific hazard ratio $\alpha_{01;i}(t; Z_i = 1)/\alpha_{01;i}(t; Z_i = 0)$ is reasonably close to the true value of 0.825. The 95% confidence interval includes 1, which is not unexpected, as most of the observed events are of the competing type,

```
> table(x$status)

   0    1    2
 160  263 1077
```

and the hazard ratio 0.825 is somewhat close to 1. Next, we analyse the competing hazard:

```
> summary(coxph(Surv(TandC,status==2)~Z,data=x))

  n= 1500

      coef exp(coef) se(coef)      z Pr(>|z|)
Z -1.61711   0.19847  0.07332 -22.06   <2e-16

  exp(coef) exp(-coef) lower .95 upper .95
Z    0.1985      5.039    0.1719    0.2291
```

The analysis is in good agreement with the theoretical cause-specific hazard ratio $\alpha_{02;i}(t; Z_i = 1)/\alpha_{02;i}(t; Z_i = 0) = 0.2$.

The partial likelihoods attached to the above cause-specific hazards analyses are $L(\beta_{01})$ and $L(\beta_{02})$ of Equation (5.17). As discussed following (5.17), we get the correct maximum likelihood estimate $\widehat{\beta} = (\widehat{\beta}_{01}, \widehat{\beta}_{02})$ by separately maximizing the $L(\beta_{0j})$s, if we assume separate models for the α_{0j}s. Our analyses above justify this assumption based on the simulated data, which is in line with the model specifications (5.26)–(5.27).

Analysis of duplicated data set We may also directly maximize the partial likelihood $L(\beta)$ of (5.9). This would also allow us to model a common effect of a covariate on both α_{0j}s. The trick is to duplicate the data as many times as there are competing events, but to leave the number of observed events unchanged (Lunn and McNeil, 1995). In our situation, an individual will enter the new data frame twice: one line corresponding to the event type 1 of interest, and the other line corresponding to the competing event. We also need a new status variable:

```
> x1 <- rbind(x,x)
> x1$eventtype <- c(rep("interest",1500), rep("competing", 1500))
> x1$newstat <- as.numeric(c(x$status == 1, x$status == 2))
```

These are the lines x1[x1$id == 1,] for individual 1 in the new data frame:

```
> a <- x1[x1$id == 1,]
> a

       id        T X.T Z       C   TandC status eventtype newstat
1       1 8.688352   2 0 89.5042 8.688352      2  interest       0
1501    1 8.688352   2 0 89.5042 8.688352      2 competing       1
```

Individual 1 has been observed to experience the competing event. Therefore, the new status indicator newstat is 1 in the line where eventtype is
competing. The entries of newstat would have been reversed if individual 1
had been observed to experience the event of interest. If individual 1 had
been censored, both entries of newstat would have equalled 0. Readers may
check this comparing respective entries of the data frames x and x1. We may
now fit a Cox model using the extended data frame and the new status variable newstat:

```
> summary(coxph(Surv(TandC, newstat != 0) ~ Z, data = x1))

Call:
coxph(formula = Surv(TandC, newstat != 0) ~ Z, data = x1)


  n= 3000


      coef exp(coef) se(coef)       z Pr(>|z|)
Z -1.28482   0.27670  0.06405 -20.06   <2e-16


    exp(coef) exp(-coef) lower .95 upper .95
Z     0.2767      3.614    0.2441    0.3137


Rsquare= 0.129    (max possible= 0.998 )
Likelihood ratio test= 415.1  on 1 df,    p=0
Wald test             = 402.4  on 1 df,    p=0
Score (logrank) test  = 442.1  on 1 df,    p=0
```

The analysis is identical to the previous first-event analysis, not distinguishing
between the two competing event types. The only differences are the values
reported for Rsquare, which we do not consider in this book for the reasons
given above, and the value n= 3000 instead of n= 1500, which reflects that
the new data frame includes two lines for every individual. The analyses are
identical, although the partial likelihoods attached to each call to coxph are
not. However, they only differ in a constant factor:

This is easily seen from the partial likelihood $L(\beta)$ of (5.9), bearing in mind
that we have *not* changed the number of observed events. As a consequence,
the number of factors in the original likelihood (attached to the analysis of
the data frame x) and in the new likelihood (attached to the analysis of the
data frame x1) are identical. Consider the factor for individual i, which was
observed to fail from event type j at time t. The corresponding factor is

$$\frac{\exp\left(\beta \cdot Z_{0j;i}\right)}{S_{0j}^{(0)}(\beta, t)} \quad \text{with} \quad S_{0j}^{(0)}(\beta, t) = \sum_{l=1}^{1500} \exp\left(\beta \cdot Z_{0j;l}\right) \cdot Y_{0;l}(t), \qquad (5.31)$$

where β and $Z_{0j;i}$ are potentially more complex than in our present simulation
example and are as explained in Section 5.1. The numerator does not change
in the new likelihood, but the denominator becomes

$$\sum_{l=1}^{1500} \exp\left(\beta \cdot Z_{0j;l}\right) \cdot Y_{0;l}(t) + \exp\left(\beta \cdot Z_{0j;l}\right) \cdot Y_{0;l}(t) = 2 \cdot S_{0j}^{(0)}(\beta, t),$$

as every individual is represented in x1 twice. As a consequence, the new likelihood is $0.5^{N_{0.}(\tau)}$ times the old likelihood, where τ is the largest observed event time (cf. page 95), and $N_{0.}(\tau)$ is the number of observed events. This does not change the estimate $\widehat{\beta}$. As seen in the example above, the variance estimates do not change either. As explained following (5.15), the variance estimates are based on the second-order partial derivatives of the log partial likelihood. Hence, the factor $0.5^{N_{0.}(\tau)}$ becomes a summand $N_{0.}(\tau) \cdot \ln(0.5)$ and then vanishes when taking derivatives.

We may check this by looking at

```
> coxph(Surv(TandC, status != 0) ~ Z, data = x)$loglik[2]
```

```
[1] -8259.807
```

which returns the values of the log partial likelihood evaluated at $\widehat{\beta}$. We compare this with

```
> coxph(Surv(TandC, newstat != 0) ~ Z,data = x1)$loglik[2]
```

```
[1] -9188.625
```

```
> - sum(x$T <= cens) * log(0.5)
```

```
[1] 928.8172
```

However, this computational subtlety is not useful in the competing risks analyses to follow, and it is desirable to directly maximize the partial likelihood $L(\beta)$ of (5.9). For this, we need to ensure that the product in (5.9) is stratified for the transition type $0 \to j$, $j = 1, 2$, i.e., the '$\prod_{j=1}^{2}$'-part in $L(\beta)$. This is achieved by including a strata(eventtype)-statement in the call to coxph:

```
> summary(coxph(Surv(TandC, newstat != 0) ~ Z +
+                strata(eventtype), data = x1))

  n= 3000

        coef exp(coef) se(coef)        z Pr(>|z|)
Z -1.28482    0.27670  0.06405 -20.06    <2e-16

    exp(coef) exp(-coef) lower .95 upper .95
Z     0.2767      3.614     0.2441     0.3137
```

This leads again to the now well-known analysis. The strata(eventtype)-statement informs coxph that we assume different baseline hazards $\alpha_{01;0}(t) \neq \alpha_{02;0}(t)$. This is essential for the competing risks analyses to follow. So far, all analyses which did not distinguish between the competing event types have been identical, because they all assumed a common effect of Z on the all-cause hazard, which follows a proportional hazards model. In addition, we note that a strata-statement is also used in coxph-analyses for single endpoint survival situations to account for patients coming from disjoint groups rather than being exposed to different endpoint types. Examples include clinical multi-center trials where patients from different centers are assumed to have different baseline survival hazards; see for example Therneau and Grambsch (2000, Section 3.2) and Kalbfleisch and Prentice (2002, Section 4.4). Such analyses typically do not use an extended data frame like xl.

The previous analysis using the extended data frame xl assumed a common effect on both cause-specific hazards. In order to model different effects on the α_{0j}s, we need to code cause-specific covariates (cf. Equation (5.3) and the discussion following it).

```
> xl$Z.01 <- xl$Z * (xl$eventtype == "interest")
> xl$Z.02 <- xl$Z * (xl$eventtype == "competing")
```

Readers may check for exemplary individuals that this coding exactly mirrors our derivation of cause-specific covariate vectors as explained following Equation (5.3). E.g., these are the respective entries for individual 754:

```
> xl[xl$id == 754, c('id', 'Z', 'Z.01', 'Z.02', 'eventtype')]

       id Z Z.01 Z.02 eventtype
754   754 1    1    0  interest
2254  754 1    0    1  competing
```

The entries Z hold the individual covariate value $Z_{754} = 1$. The entries Z.01 hold the individual cause-specific covariate vector $Z_{01;754} = (1,0)^\top$. Analogously, Z.02 corresponds to $Z_{02;754} = (0,1)^\top$.

We are now prepared to study the different effects of Z on the cause-specific hazards. In our call to coxph, we specify that we assume different baseline hazards $\alpha_{01;0}(t)$ and $\alpha_{02;0}(t)$ through the strata(eventtype)-statement.

```
> summary(coxph(Surv(TandC, newstat != 0) ~ Z.01 + Z.02 +
+               strata(eventtype), data = xl))

  n= 3000

          coef exp(coef) se(coef)      z Pr(>|z|)
Z.01 -0.16825   0.84514  0.12744  -1.32    0.187
Z.02 -1.61711   0.19847  0.07332 -22.06   <2e-16


     exp(coef) exp(-coef) lower .95 upper .95
Z.01    0.8451      1.183    0.6583    1.0849
Z.02    0.1985      5.039    0.1719    0.2291
```

The analysis is identical to those obtained from fitting two separate Cox models, except for a slightly different rounding in the output. Readers may check that including the strata(eventtype)-statement is now essential and that omitting it leads to erroneous results.

So far, we have not gained anything from extending the original data frame to twice the number of lines. The advantage of the extended data frame is that it allows us to model both cause-specific and common effects on the cause-specific hazards. We later also use the extended data frame when making predictions, using the mstate package. To illustrate the first point, we generate a random variable which has nothing to do with the way the data were simulated and should therefore have a common regression coefficient 0:

```
> cv <- round(0.5 * rnorm(length(x$id),1))
> xl$cv <- rep(cv, 2)
```

As before, let us look at the respective entries for individual 754:

```
      id Z Z.01 Z.02 cv eventtype
754  754 1    1    0  1  interest
2254 754 1    0    1  1 competing
```

The cause-specific covariate vectors are now $(1, 0, 1)^\top$ for the event type 1 of interest and $(0, 1, 1)^\top$ for the competing event type 2. Below is the output of the analysis of the cause-specific hazard for event type 1,

```
> summary(coxph(Surv(TandC, status == 1) ~ Z + cv, data = x))

  n= 1500


       coef exp(coef) se(coef)      z Pr(>|z|)
Z  -0.16873   0.84473  0.12746 -1.324    0.186
cv -0.02403   0.97626  0.10623 -0.226    0.821


   exp(coef) exp(-coef) lower .95 upper .95
Z     0.8447      1.184    0.6580     1.084
cv    0.9763      1.024    0.7928     1.202
```

of the analysis of the competing cause-specific hazard $\alpha_{02}(t)$,

```
> summary(coxph(Surv(TandC, status == 2) ~ Z + cv,data = x))

  n= 1500


      coef exp(coef) se(coef)       z Pr(>|z|)
Z  -1.61285   0.19932  0.07334 -21.991   <2e-16
cv  0.06334   1.06539  0.05220   1.214    0.225


   exp(coef) exp(-coef) lower .95 upper .95
Z     0.1993     5.0171    0.1726    0.2301
cv    1.0654     0.9386    0.9618    1.1802
```

and of the simultaneous analysis of both hazards,

```
> summary(coxph(Surv(TandC, newstat != 0) ~ Z.01 + Z.02 + cv +
+                strata(eventtype), data = x1))
```

  n= 3000

```
          coef exp(coef) se(coef)        z Pr(>|z|)
Z.01 -0.16731   0.84594  0.12744  -1.313    0.189
Z.02 -1.61400   0.19909  0.07334 -22.007   <2e-16
cv    0.04637   1.04746  0.04683   0.990    0.322


     exp(coef) exp(-coef) lower .95 upper .95
Z.01    0.8459     1.1821    0.6590    1.0860
Z.02    0.1991     5.0229    0.1724    0.2299
cv      1.0475     0.9547    0.9556    1.1482
```

all illustrate that all analyses reasonably well detect that cv has no effect. The 'precision', however, in terms of the variance estimate has increased with more events present in the analysis.

The common effect of cv on both cause-specific hazards was, of course, due to the toy example character of the generation of cv. If, in an actual data analysis, the data suggest that we may assume a common effect for some covariate, this may be desirable in terms of model parsimony. However, assuming a common effect may also impose difficulties in the interpretation. Andersen et al. give a pertinent example (Andersen et al., 1993, p. 494). More recently, Glynn et al. (2009) studied proportional cause-specific hazards models for the development of different types of cataract; the authors found cataract type-specific effects of diabetes and body mass index, but a common effect for other covariates such as smoking. Going back to the motivation of the present simulation setting, it may be hard to imagine a covariate that displays the same effect on the cause-specific infection hazard and on the cause-specific end-of-neutropenia hazard.

Breslow estimators, Nelson-Aalen estimators and interpretation We estimate the cumulative cause-specific baseline hazards $A_{0j;0}(t) = \int_0^t \alpha_{0j;0}(u)du$ using the simulated data, $j = 1, 2$. Figure 5.6 shows a customized plot of the Breslow estimators (5.18).

As we have assumed completely separate models for the α_{0j}s, the Breslow estimators may be obtained from the separate Cox models,

```
> a01.0 <- basehaz(coxph(Surv(TandC, status == 1) ~ Z,
+          data = x), centered = FALSE)
> a02.0 <- basehaz(coxph(Surv(TandC, status == 2) ~ Z,
+          data = x), centered = FALSE)
```

with self-explanatory entries time and hazard. The statement centered = FALSE ensures that the Breslow estimator of the baseline hazard is computed.

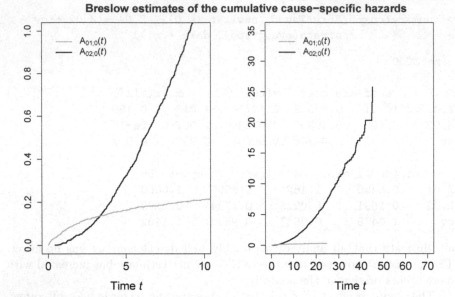

Fig. 5.6. *Simulated data.* Breslow estimators of the cumulative cause-specific baseline hazards $A_{0j;0}(t)$, $j = 1, 2$. The left plot displays the early time interval $[0, 10]$ at a different scale.

If we omit this statement, the predictor (5.19) is computed at the empirical covariate mean. We may also obtain both Breslow estimators at once using the extended data frame; this also works if we assume some covariate to have a common effect on both α_{0j}s:

```
> basehaz(coxph(Surv(TandC, newstat != 0) ~ Z.01 + Z.02 +
+               strata(eventtype), data = x1), centered = FALSE)
```

The return value now also has an entry `strata` in order to tell the Breslow estimators $\widehat{A}_{01;0}(t)$ and $\widehat{A}_{02;0}(t)$ apart. Readers may check that the Breslow estimates in Figure 5.6 are in very good agreement with both the Nelson-Aalen estimates, which may be computed as described in Section 4.2, and the true cumulative cause-specific hazards.

The *interpretation* of Figure 5.6 is that if an event occurs very early in the baseline group, it will likely be of type 1; otherwise, it will likely be of the competing type 2. In our preceding discussion of the model specification, we have emphasized the crucial role played by the baseline hazards. This is summarized in Figure 5.7, where the Nelson-Aalen estimates of the cumulative cause-specific hazards are plotted for both groups in the early time interval $[0, 10]$. The interpretation of Figure 5.7 is that individuals with $Z_i = 1$ stay longer in the initial state of the competing risks process. The prolonged stay is mostly due to a pronounced reducing effect on the competing cause-

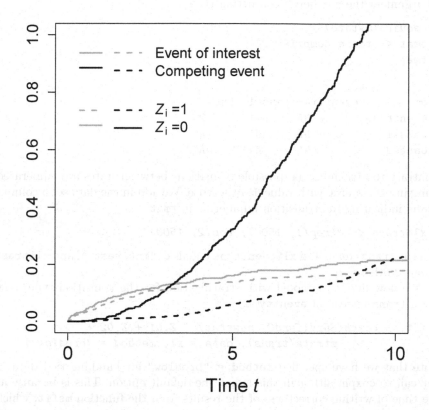

Fig. 5.7. *Simulated data.* Nelson-Aalen estimators of the cumulative cause-specific hazards $A_{0j;i}(t; Z_i)$, $j = 1, 2$.

specific hazard, which is eventually also the major hazard (cf. also Figure 5.6). During this prolonged time, individuals are exposed to an only slightly reduced cause-specific hazard $\alpha_{01}(t)$. Therefore, we eventually see more type 1 events for individuals with $Z_i = 1$, but the increase comes a bit delayed.

Finally, this effect is displayed in Figures 5.8 and 5.9, which are the empirical analogue of Figures 5.3 and 5.4. Figures 5.8 and 5.9 show both the Aalen-Johansen estimates of the cumulative incidence functions, stratified for covariate value Z_i, and their model based counterparts (5.23). How to obtain the latter in R is discussed next.

Prediction of the cumulative incidence functions The predicted cumulative incidence functions (5.23) can be computed using the mstate package (de Wreede et al., 2010, 2011). For this, we need the extended data xl computed in Section 5.2.2. Similar to `etm` and `mvna`, we define a matrix indicating the possible transitions using the convenience function `trans.comprisk`, which takes as an argument the number of competing risks,

```
> require(mstate)
> tmat <- trans.comprisk(2)
> tmat

            to
from           eventfree cause1 cause2
  eventfree          NA      1      2
  cause1             NA     NA     NA
  cause2             NA     NA     NA
```

Contrary to `etm` and `mvna`, possible transitions between states are numerated using integer values, with value `NA` otherwise. We add in the data set a column `trans` indicating the transition number as in `tmat`

```
> xl$trans <- c(rep(1, 1500), rep(2, 1500))
```

I.e., `xl$trans` equals 1, if `xl$eventtype` equals `c("interest")`, and `xl$trans` equals 2 otherwise.

We now fit a Cox model with stratification for the transition type, now using `trans` instead of `eventtype`.

```
> fit <- coxph(Surv(TandC, newstat) ~ Z.01 + Z.02 +
+              strata(trans), data = xl, method = "breslow")
```

Note that we have specified `method = "breslow"` for handling tied data in our call to `coxph`, although this is not the default option. This is because at the time of writing correctness of the results from the function `msfit`, which we use below, had been checked for this case.

In order to obtain predicted cumulative incidence functions for a specific value of the covariate Z, we create a new data frame containing the transition specific covariate values for each of the possible transitions. Such a data frame should hold the covariates of one individual.

```
> ## An individual with Z=0
> newdat.z0 <- data.frame(Z.01 = c(0, 0), Z.02 = c(0, 0),
+                         strata = c(1, 2))
> ## An individual with Z=1
> newdat.z1 <- data.frame(Z.01 = c(1, 0), Z.02 = c(0, 1),
+                         strata = c(1, 2))
```

The new data frames must contain the same names for the transition specific covariates as those used for fitting the Cox model, plus a column `strata` indicating the corresponding transition number as in `tmat`. We then apply the `msfit` function to obtain predictive cumulative transition hazards as defined in (5.19). This function takes as arguments the fitted Cox model, the new data set for which we wish to make prediction, and `tmat`, the matrix specifying the possible transitions.

```
> msf.z0 <- msfit(fit, newdat.z0, trans = tmat)
> msf.z1 <- msfit(fit, newdat.z1, trans = tmat)
```

msfit returns a list with entries `Haz` for the predicted cumulative hazards, `varHaz` containing estimated variances and covariances, and `trans` containing the matrix `tmat`. The entry `varHaz` will be omitted if we specify `variance=FALSE` in our call to `msfit`. The entry `Haz` is a data frame with entries `time`, `Haz`, and `trans` containing the event times, the predicted cumulative hazards, and the corresponding transition number as in `tmat`. The entry `varHaz` has an analogous structure.

The predicted cumulative incidences are then finally computed using the `probtrans` function, which takes as arguments an `msfit` object and the time at which prediction starts, in our case at time 0.

```
> pt.z0 <- probtrans(msf.z0, 0)[[1]]
> pt.z1 <- probtrans(msf.z1, 0)[[1]]
```

We select only item `[[1]]` from the output of the `probtrans` function, which contains the predicted transition probabilities out of the initial state. Only these are of interest in a competing risks setting. In our example, `probtrans` returns a list with two additional items containing the transition probabilities out of the absorbing states. These are degenerated in the sense that they are always equal to zero for the probability to move out of an absorbing state and always equal to one for the probability to stay in an absorbing state. Also note that `probtrans` will return an additional item containing covariances of the predicted transition probabilities, if we specify `covariance=TRUE` in our call to `probtrans`.

Running `probtrans` may produce warnings.

```
Warning! Negative diagonal elements of (I+dA);
the estimate may not be meaningful.
```

The warning indicates increments (5.22), which are larger than 1. A quick computation shows that this will typically only happen for small risk sets and/or extreme covariate values.

E.g., `probtrans(msf.z0, 0)[[1]]` is a data frame with the event times in entry `time`, the predicted survival function in entry `pstate1` and the predicted cumulative incidence functions in entries `pstate2` and `pstate3`. Here, the numbering of `pstate1` through `pstate3` corresponds to the sequence with which they appear in `tmat`, i.e.,

```
> dimnames(tmat)

$from
[1] "eventfree" "cause1"    "cause2"
```

```
$to
[1] "eventfree" "cause1"    "cause2"
```

In other words, `pstate1` contains the estimates $\widehat{P}(T > t \mid z)$ as in (5.21), `pstate2` contains the estimates $\widehat{P}(T \leq t, X_T = 1 \mid z)$ as in (5.23) for $j = 1$, and `pstate3` contains the estimates $\widehat{P}(T \leq t, X_T = 2 \mid z)$ as in (5.23) for $j = 2$. The corresponding standard errors are in the entries `se1` through `se3`. Computing the standard errors may be suppressed by specifying `variance=FALSE` in the call `probtrans`.

Figures 5.8 and 5.9 display the Aalen-Johansen estimates and the predicted cumulative incidence functions for both values of the covariate Z. The x-axes in these figures are chosen as in Figure 5.4.

The curves are in close agreement.

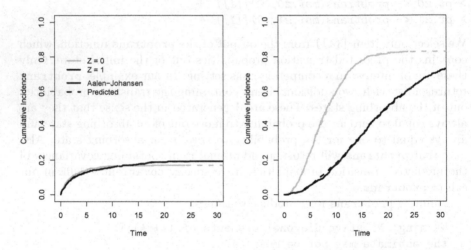

Fig. 5.8. *Simulated data.* Aalen Johansen estimates of the cumulative incidence functions for $X_T = 1$ (left plot) and $X_T = 2$ (right plot), along with the predicted cumulative incidence functions in dashed lines. See Figure 5.9 for a presentation of the left plot with a different scale for the y-axis.

Analysis of hospital data: Impact of pneumonia status on admission on intensive care unit mortality

We consider the hospital data that we have analysed in Section 4.3 in a nonparametric fashion. Recall, in particular, from the Nelson-Aalen estimators of the cumulative cause-specific hazards in Figure 4.7 that pneumonia

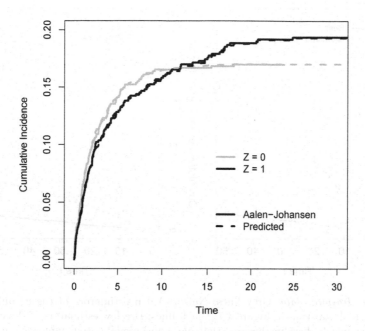

Fig. 5.9. *Simulated data.* Aalen Johansen estimates of the cumulative incidence functions for $X_T = 1$ along with the predicted cumulative incidence functions in dashed lines as in Figure 5.8, but with a different scale for the y-axis. This figure is an empirical analogue of Figure 5.4.

on admission increases hospital mortality through a decreasing effect on the alive discharge hazard, whereas the hospital death hazard is left essentially unchanged. The aim of the present analysis is to reinvestigate this finding via proportional cause-specific hazards models.

We use the data frame `my.sir.data` that has been generated at the beginning of Section 4.3. Figure 4.7 suggests that pneumonia has different effects on the cause-specific hazards. We therefore simply fit two different Cox models as explained earlier. This is the output of the analysis of the cause-specific hazard of interest for hospital death,

```
> summary(coxph(Surv(time, to == 1) ~ pneu, data = my.sir.data))

Call:
coxph(formula = Surv(time, to == 1) ~ pneu, data = my.sir.data)

  n= 747
```

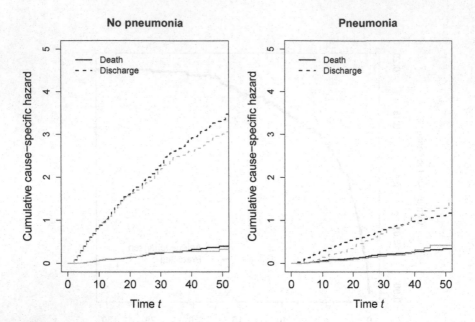

Fig. 5.10. *Hospital data.* Grey lines: Nelson-Aalen estimators of the cumulative cause-specific hazards as in Figure 4.7. Black lines: Breslow estimators of the cumulative cause-specific baseline hazards (left plot) and model-based cumulative hazard estimators (5.19) for patients with pneumonia (right plot).

```
            coef exp(coef) se(coef)       z Pr(>|z|)
pneu -0.1622    0.8503   0.2678 -0.606    0.545

        exp(coef) exp(-coef) lower .95 upper .95
pneu    0.8503      1.176      0.503      1.437

Rsquare= 0.001    (max possible= 0.651 )
Likelihood ratio test= 0.37  on 1 df,    p=0.5407
Wald test                = 0.37  on 1 df,    p=0.5448
Score (logrank) test = 0.37  on 1 df,    p=0.5445
```

together with the output of the analysis of the competing cause-specific hazard for discharge,

```
> summary(coxph(Surv(time, to == 2) ~ pneu, data = my.sir.data))

Call:
coxph(formula = Surv(time, to == 2) ~ pneu, data = my.sir.data)

  n= 747
```

```
          coef exp(coef) se(coef)        z Pr(>|z|)
pneu  -1.0901     0.3362   0.1299 -8.391    <2e-16


       exp(coef) exp(-coef) lower .95 upper .95
pneu     0.3362      2.974    0.2606    0.4337


Rsquare= 0.116    (max possible= 1 )
Likelihood ratio test= 91.7  on 1 df,    p=0
Wald test           = 70.4  on 1 df,    p=0
Score (logrank) test = 77.09  on 1 df,    p=0
```

The proportional cause-specific hazards analyses are in good agreement with our previous findings.

The R output above also re-emphasizes two important aspects in the analysis of competing risks data: first, all cause-specific hazards should be analysed. We should by no means conclude from a cause-specific death hazard ratio of 0.85 with 95% confidence interval [0.503, 1.437] that pneumonia appears to have no impact on hospital death. Second, the cause-specific hazard ratios displayed above make no statement about the magnitude of the cause-specific baseline hazards. This is quite unlike our initial analysis based on the Nelson-Aalen estimators in Figure 4.7.

Figure 5.10 shows a customized plot of the Nelson-Aalen estimators together with the Breslow estimators (for no pneumonia) and the model-based cumulative hazard estimators (5.19) (for pneumonia on admission). We have restricted the time axis to [0, 50], because most events happen in that time interval judged from Figure 4.9. (In fact, one finds that time 47 is the empirical 95% quantile, estimated following Andersen et al. (1993, Section IV.3.4).) We find that all hospital death-specific curves are in good agreement, as are the baseline estimators for the cumulative discharge hazard. However, the respective estimators of the cumulative discharge hazard for pneumonia patients do not agree quite as well, indicating that the effect of pneumonia on the discharge hazard may not follow a proportional cause-specific hazards model. Figure 5.10 suggests that the estimated cause-specific discharge hazard ratio of 0.336 with 95% confidence interval [0.261, 0.434] reports a time-averaged effect of pneumonia on the discharge hazard. The notion of a a time-averaged effect is explained in more detail in Section 5.4.

Analysis of pregnancy outcome data: Impact of coumarin derivatives on spontaneous and induced abortion

Finally, we briefly illustrate how to fit proportional cause-specific hazards models with left-truncated data. We consider the data set abortion as in Section 4.4. Recall from Section 4.2 that both the left-truncation times and the event times (or censoring times) are now being passed to Surv. The following code analyses the impact of coumarin derivatives on the cause-specific hazards of induced abortion and spontaneous abortion, respectively.

```
> coxph(Surv(entry,exit, cause == 1) ~ group,
+         data = abortion)# induced

Call:
coxph(formula = Surv(entry, exit, cause == 1) ~ group,
                                    data = abortion)


      coef exp(coef) se(coef)    z p
group 2.36      10.6    0.278 8.49 0

Likelihood ratio test=74.2  on 1 df, p=0  n= 1186

> coxph(Surv(entry,exit, cause == 3) ~ group,
+         data = abortion)# spontaneous

Call:
coxph(formula = Surv(entry, exit, cause == 3) ~ group,
                                    data = abortion)


      coef exp(coef) se(coef)    z       p
group 1.23      3.42    0.196 6.27 3.6e-10

Likelihood ratio test=34.3  on 1 df, p=4.82e-09  n= 1186
```

The respective analysis for live birth is

```
> coxph(Surv(entry,exit, cause == 2) ~ group, data = abortion)
Call:
coxph(formula = Surv(entry, exit, cause == 2) ~ group,
                                    data = abortion)


       coef exp(coef) se(coef)    z     p
group 0.223      1.25    0.109 2.04 0.042

Likelihood ratio test=3.9  on 1 df, p=0.0482  n= 1186
```

The analyses are in line with the findings in Section 4.4, where abortion proportions were seen to be increased for women exposed to coumarin derivatives.

5.3 Proportional subdistribution hazards model

We consider the proportional subdistribution hazards model (5.7) with individual covariate vectors Z_i. The model is custom made to analyse one cumulative incidence function of interest. As a consequence, there is only one subdistribution hazard, which is assumed to follow a Cox model. In contrast, there were two cause-specific hazards in the preceding Section 5.2, both of which were assumed to follow a Cox model. The subdistribution framework is derived from the standard competing risks multistate model in Section 5.3.1. We introduce estimation and prediction in Section 5.3.2. We apply these methods in Section 5.3.3 to data which have been previously analysed in Section 5.2.2. The relative merits of the approach based on cause-specific hazards and the one based on the subdistribution hazard, respectively, are discussed.

Although custom made to model one cumulative incidence function only, the proportional subdistribution hazards model is often used to analyse all cumulative incidence functions. Section 5.3.4 explains that this approach presents conceptual problems, which requires us to interpret such analyses in terms of a time-averaged effect.

Technical difficulties discussed in Section 5.3.2 have led to the analysis of subdistribution hazards having mainly been developed for right-censored data only. We briefly cover the case of left-truncation in Section 5.3.5. Finally, simulating data that follow a proportional subdistribution hazards model is described in Section 5.3.6, but the cause-specific hazards that generate such data are found to follow rather involved models.

5.3.1 The subdistribution process

The analysis of competing risks data requires analysing all of the cause-specific hazards. As explained in the Introduction Section 5.1, the aim of a subdistribution hazards analysis is to provide for a single analysis of a different quantity, the subdistribution hazard, that allows for direct interpretation of the results in terms of one cumulative incidence function, just as we could interpret the single analysis of an all-cause hazard in Section 5.2.2 in terms of the waiting time distribution in the initial state.

In the brief Introduction 5.1, we defined the subdistribution hazard $\lambda(t)$ rather technically, requiring that the cumulative incidence function $P(T \leq t, X_T = 1)$ for event type 1 equals $1 - \exp(-\int_0^t \lambda(u)du)$. This mimics the usual one minus survival function formula, but with the subdistribution hazard replacing the all-cause hazard. In this section, we approach the subdistribution framework from a process/multistate point of view, which we feel is more intuitive. It also has the advantage of clearly displaying the connection to the original competing risks model and it is easily generalized to more complex multistate models with competing absorbing states; see Section 11.2.2.

Consider the original competing risks process $(X_t)_{t \geq 0}$ of Figure 3.1, redisplayed in Figure 5.11 (left). The event time T is the waiting time of $(X_t)_{t \geq 0}$ in the initial state 0 (i.e., until occurrence of any first event). The idea of the subdistribution framework is to consider a subdistribution time ϑ until occurrence of the event 1 of interest. This needs to account for the fact that occurrence of a first event can be of the competing type 2. The right way to do

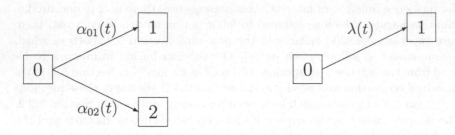

Original competing risks process $(X_t)_{t \geq 0}$ Subdistribution process $(\xi_t)_{t \geq 0}$

Fig. 5.11. Left: original competing risks process $(X_t)_{t \geq 0}$ of Figure 3.1 with cause-specific hazards $\alpha_{0j}(t)$, $j = 1, 2$. Right: subdistribution process $(\xi_t)_{t \geq 0}$ with subdistribution hazard $\lambda(t)$. $(\xi_t)_{t \geq 0}$ is derived from $(X_t)_{t \geq 0}$ by suitably stopping the original process. This leads to different interpretations of being in the initial state for each process.

this is by *stopping* the original competing risks process just prior to T, if the process moves into the competing event state 2. Hence, the subdistribution process $(\xi_t)_{t \geq 0}$ of Figure 5.11 (right) is defined as

$$\xi(t) := \mathbf{1}(X(t) \neq 2) \cdot X(t) + \mathbf{1}(X(t) = 2) \cdot X(T-). \qquad (5.32)$$

In words, X_t and ξ_t are equal, as long as the original process does not move into the competing event state 2. If $(X_t)_{t \geq 0}$ moves into state 2 at time T, individuals 'get stuck' in the initial state of the subdistribution process: ξ_t remains equal to 0 for all $t \geq T$, if $X_T = 2$. As a consequence, the interpretation of being in the respective initial states 0 changes: $X(t) = 0$ has the interpretation 'no event by time t', and $\xi(t) = 0$ means 'no type 1 event by time t'. Following from this, we may now define the time until occurrence of the event 1 of interest as

$$\vartheta := \inf\{t > 0 \,|\, \xi_t \neq 0\}. \qquad (5.33)$$

The set on the right hand side of (5.33) can be empty. If $X_T = 2$, ξ_t equals 0 for all times t and there is no smallest time such that the subdistribution process is unequal to 0. In such a case, ϑ is the infimum of an empty set, which (as is usual) is defined to be infinity: $\vartheta = \infty$, if $X_T = 2$. The interpretation of this is that there is no finite time at which the subdistribution process enters the event state 1, if $X_T = 2$. However, if the original competing risks process enters state 1, so does the subdistribution process, and we have $T = \vartheta$. In summary,

$$\vartheta = \begin{cases} T, & \text{if } X_T = 1 \\ \\ \infty, & \text{if } X_T = 2. \end{cases} \qquad (5.34)$$

As a consequence, the distribution function of the subdistribution failure time ϑ equals the cumulative incidence function for event type 1,

$$P(\vartheta \leq t) = P(T \leq t, X_T = 1) \text{ for all } t \in [0, \infty), \qquad (5.35)$$

and $P(\vartheta = \infty) = P(X_T = 2)$. The subdistribution hazard is now defined as the hazard 'attached to ϑ',

$$\lambda(t)dt := P(\vartheta \in dt \mid \vartheta \geq t). \qquad (5.36)$$

Note that this definition parallels the definition of the all-cause hazard in (3.7) but with ϑ in place of T. As a consequence,

$$P(\vartheta \leq t) = 1 - \exp(-\int_0^t \lambda(u) \, du) \qquad (5.37)$$

(cf. (3.10)). Because of Equations (5.35) and (5.37), the subdistribution framework reestablishes a one-to-one correspondence between subdistribution hazard and cumulative incidence function, which is otherwise an involved function of both cause-specific hazards $\alpha_{01}(t)$ and $\alpha_{02}(t)$ (cf. (3.11)):

$$1 - \exp(-\int_0^t \lambda(u) \, du) = \int_0^t \exp\left(-\int_0^u \alpha_{01}(v) + \alpha_{02}(v)dv\right) \alpha_{01}(u) \, du.$$
$$(5.38)$$

The idea of the proportional subdistribution hazards model suggested by Fine and Gray (1999) and discussed below is to fit a Cox model to the subdistribution hazard, which has a direct probability interpretation in terms of the cumulative incidence function $P(T \leq t, X_T = 1)$ for event type 1 because of Equations (5.35) and (5.37). However, fitting the model in practice has to deal with certain technical difficulties implied by ϑ having mass at infinity. This is discussed below.

Equation (5.38) implies that

$$\alpha_{01}(t) = \left(1 + \frac{P(T \leq t, X_T = 2)}{P(T > t)}\right) \cdot \lambda(t). \qquad (5.39)$$

Equation (5.39) has a number of important consequences: as already mentioned in the Introduction in Section 5.1, $\lambda(t)$ is weighted down as compared to $\alpha_{01}(t)$, ensuring that $P(X_T = 1) = 1 - \lim_{t \to \infty} \exp\left(-\int_0^t \lambda(u)du\right) = 1 - P(X_T = 2)$ (cf. (5.6)). As the weighting is time-dependent, assuming a proportional hazards model for $\alpha_{01}(t)$ precludes that $\lambda(t)$ follows a proportional subdistribution hazards model, and vice versa. We address the consequences of such potentially misspecified models in the data analyses below and in Section 5.4. The weighting also depends on occurrence of the competing events, reflecting that a cumulative incidence function depends on all cause-specific hazards. (Latouche (2004) suggested a nice graphical presentation of this fact.)

Furthermore, we may compute the subdistribution hazard-based on knowledge of the original process, as should be expected from our derivation of the subdistribution process. Hence, we should be able to make a decent guess on the results of a subdistribution hazards analysis, if we have previously analysed all cause-specific hazards, including the baseline hazards. This is also addressed in the data analyses.

We note that Gray (1988) first suggested considering subdistribution failure times when analysing right-censored competing risks data. Gray called the subdistribution failure time 'improper', because its distribution has mass at infinity. He also noted that the counting process approach (Andersen et al., 1993) does in principle allow for such random variables. Next, in Section 5.3.1, we consider fitting a proportional subdistribution hazards model, together with certain technical difficulties connected to the subdistribution failure time being 'improper'. We also note that defining an event time as infinity, if the event of interest does not occur, is natural from a probabilistic modelling point of view (cf. Shiryaev (1995, Chapter 10) for a textbook example).

5.3.2 Estimation

Estimation and prediction for the proportional subdistribution hazards model is, in spirit, similar to a standard Cox model; see Section 5.2.1. We explain that the subdistribution process, however, poses a problem in that the attached risk set may be unknown. Solutions have mainly been developed for right-censored data and require explicit modelling of the censoring distribution. In practice, random censorship is typically assumed. We describe the available approaches that either assume administrative censoring or tackle the problem of unknown risk sets using inverse probability of censoring weighting or multiple imputation. At the time of writing, estimation in the presence of left-truncation was a topic of ongoing research, and a further discussion of this issue is deferred to Section 5.3.5.

Counting process of observed subdistribution events, subdistribution at-risk process, and technical difficulties

In principle, the proportional subdistribution hazards model is 'just' a proportional hazards model. Under a data structure as in Section 5.2.1, we may invoke the respective estimation techniques, simplified to the extent that we only have one subdistribution hazard instead of two cause-specific hazards. In order to do this, we need to consider both a counting process of observed events and an at-risk process based on the subdistribution process derived in Section 5.3.1. We define the individual counting process

$$N_i^\star(t) := \mathbf{1}(\vartheta_i \wedge C_i \leq t, L_i < \vartheta_i \leq C_i) \tag{5.40}$$

of an observed subdistribution event and the individual subdistribution at-risk process

$$Y_{0;i}^{\star}(t) := \mathbf{1}(L_i < t \le \vartheta_i \wedge C_i). \tag{5.41}$$

These $\star$-processes are as the usual processes in Equations (4.1) and (4.2), but with the individual subdistribution failure time ϑ_i in place of the 'real-life' failure time T_i. Also note that we have only one process $N_i^{\star}(t)$ of observed events, as the subdistribution framework is tailored to investigate type 1 events. We also write $N^{\star}(t)$ and $Y_0^{\star}(t)$ for the respective processes aggregated over all individuals i, $i = 1, 2, \ldots n$, as in Section 4.1.

Note that $N_i^{\star}(t)$ will equal 1, if and only if we have observed ϑ_i up to and including time t. In such a case, a type 1 event has been observed, and we have $\vartheta_i = T_i$. As a consequence,

$$N_i^{\star}(t) = N_{01;i}(t) \text{ and } N^{\star}(t) = N_{01}(t), \tag{5.42}$$

where $N_{01;i}(t)$ and $N_{01}(t)$ are the original type 1 counting processes (4.2) and (4.5), respectively.

The technical difficulty of the subdistribution framework, alluded to in the previous section, arises through the at-risk processes $Y_{0;i}^{\star}(t)$. Readers can easily verify that these are known for individuals under study and with no type 1, type 2 nor censoring event yet, for individuals with an observed type 1 event and for individuals, who have been right-censored. However, $Y_{0;i}^{\star}(t)$ is not known for the following individuals.

(a) Individuals who have been observed to experience a type 2 event; this is illustrated in Figure 5.12 a).
(b) Individuals who never entered the study, because their failure time was less than or equal to their left-truncation time, and who experienced a type 2 event; this is illustrated in Figure 5.12 b).

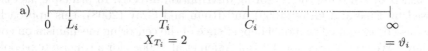

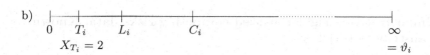

Fig. 5.12. a) Individual with an observed type 2 event. b) Individual with a type 2 event before study entry.

We discuss Figure 5.12 a) first. For such individuals, we know both T_i and that $X_{T_i} = 2$. As a consequence, we know that $\vartheta_i = \infty$ and $\vartheta_i \wedge C_i = C_i$.

However, observation may have stopped at T_i; e.g., if the individual ceases to exist afterwards, because the competing endpoint 2 is death. But even if the competing state 2 is not fatal in real life, the study protocol may have planned to collect data only until T_i (regardless of the endpoint type) or until C_i, whatever comes first. As a consequence, in general, we do not know C_i, which is a future event with respect to T_i in Figure 5.12 a). Hence, $Y_{0;i}^*(t)$ is unknown for such an individual for $t \in (T_i, C_i]$.

Next, we discuss Figure 5.12 b): the individual illustrated in this figure never enters the study, because T_i is less than L_i. This would not be problematic, if X_{T_i} were equal to 1. Here, however, the individual experiences the competing event type (i.e., $X_{T_i} = 2$) and we have again that $\vartheta_i = \infty$. As a consequence, $\vartheta_i > L_i$, i.e., we should have seen this individual in terms of the subdistribution process, and it should have been at risk (in the sense of $Y_{0;i}^*(t)$) in the time interval $(L_i, C_i]$. The problem here is twofold: because the individual has not entered the study in 'real life', we are typically not even aware of the individual nor do we know L_i and C_i.

These difficulties have been solved and implemented in R in the presence of random right-censoring (Fine and Gray, 1999; Ruan and Gray, 2008), and the situation of Figure 5.12 b) was the subject of ongoing research work at the time of writing (Zhang et al., 2009; Geskus, 2011; Zhang et al., 2011). We consider randomly right-censored competing risks data below. We briefly comment on left-truncation in Section 5.3.5.

We note that the assumption of *random* censorship is more restrictive than the assumption of *independent* censorship, which is typically made when modelling cause-specific hazards (cf. Section 2.2.2). In particular, censoring may depend on the covariates included in the model, if cause-specific hazards are analysed. Such a potential dependence is not accounted for when assuming random censorship. The reason for assuming random censoring is unknown subdistribution at-risk processes. As we show below, this technical difficulty is tackled by studying the censoring distribution directly. In principle, as discussed by Fine and Gray (1999) and Ruan and Gray (2008), this approach may also be extended to model dependence of the censoring mechanism on covariates. We refer to Section 4.2.7 of Aalen et al. (2008) for a concise textbook account on this issue.

Administratively right-censored competing risks data: Known subdistribution risk set

Using (5.34), the subdistribution at-risk process becomes

$$Y_{0;i}^*(t) = \mathbf{1}(t \le T_i \wedge C_i) + \mathbf{1}(T_i < t \le C_i, X_{T_i} = 2) \tag{5.43}$$

in the absence of left-truncation. The first summand on the right hand side of (5.43) is known, the second summand is usually unknown. An exception, where $\mathbf{1}(T_i < t \le C_i, X_{T_i} = 2)$ is known, is administrative censoring. In this case, censoring is solely due to administrative termination of the study, such that an individual's potential censoring time is known for each individual in advance. As a consequence, the technical difficulty discussed above in the context of Figure 5.12 a) disappears and we can use standard Cox software. This is illustrated in the examples below.

Because the subdistribution at-risk process is known in such a situation, it has been dubbed 'censoring complete data' in Fine and Gray (1999). This censoring mechanism is also known as (progressive, generalized) Type I censoring. In-depth discussions can be found in Andersen et al. (1993, Example III.2.3) and Klein and Moeschberger (2003, Section 3.2).

Randomly right-censored competing risks data: 'estimated' subdistribution risk set

As (5.43) is in general unknown, the idea of Fine and Gray was to replace $Y^*_{0;i}(t)$ by an 'estimated' risk set

$$\widehat{Y^*}_{0;i}(t) :=$$

$$\mathbf{1}(C_i \geq T_i \wedge t) \cdot \frac{\widehat{G}(t-)}{\widehat{G}(\{T_i \wedge C_i \wedge t\}-)} \cdot \left(\mathbf{1}(t \leq T_i) + \mathbf{1}(T_i < t, X_{T_i} = 2)\right),$$

$$(5.44)$$

where $\widehat{G}(t)$ is the Kaplan-Meier estimator of the censoring survival function $P(C > t)$ using the 'censoring event indicator' $1 - \mathbf{1}(T \leq C)$. Usage of $\widehat{G}(t)$ as in (5.44) is an application of inverse probability of censoring weighting (Robins and Rotnitzky, 1992); the weighting is applied to $\mathbf{1}(t \leq T_i) + \mathbf{1}(T_i < t, X_{T_i} = 2)$, which is the subdistribution risk set in the case of *complete* data. The idea behind estimator (5.44) is this:

First, it is computable from the observable data. The term $\mathbf{1}(C_i \geq T_i \wedge t)$ equals one, if we have knowledge of individual i's vital status just prior to t: $\mathbf{1}(C_i \geq T_i \wedge t) = 1$ for individuals with no type 1, type 2, nor censoring event in $[0, t)$ (if $t \leq T_i$) and for individuals with an observed type 1 or type 2 event in $[0, t)$ (if $T_i < t$). For such an individual, the remainder of (5.44) can be computed from the observable data. Note that the last factor on the right hand side of (5.44) equals 1, if and only if individual i's subdistribution process (5.32) is in the initial state 0.

Second, $\widehat{Y^*}_{0;i}(t)$ is asymptotically unbiased in the sense that its expectation equals $E\left(Y^*_{0;i}(t)\right)$ asymptotically. To see this, note that (5.44) equals

$$\mathbf{1}(t \leq T_i \wedge C_i) + \mathbf{1}(C_i \geq T_i)\frac{\widehat{G}(t-)}{\widehat{G}(T_i-)}\mathbf{1}(T_i < t, X_{T_i} = 2). \qquad (5.45)$$

Because of the representation of $Y^*_{0;i}(t)$ in (5.43), we only need to consider the second summand in (5.45). Now consider the expectation of this term conditional on $T_i = s, X_{T_i} = j$, i.e.,

$$E\left(\mathbf{1}(C_i \geq s)\frac{\widehat{G}(t-)}{\widehat{G}(s-)}\mathbf{1}(s < t, j = 2) \mid T_i = s, X_{T_i} = j\right)$$

$$= \mathbf{1}(s < t, j = 2) \cdot E\left(\mathbf{1}(C_i \geq s)\frac{\widehat{G}(t-)}{\widehat{G}(s-)} \mid T_i = s, X_{T_i} = j\right)$$

$$= \mathbf{1}(s < t, j = 2) \cdot P(C \geq s) \cdot E\left(\frac{\widehat{G}(t-)}{\widehat{G}(s-)} \mid T_i = s, X_{T_i} = j\right). \quad (5.46)$$

Note that the term $\widehat{G}(t-)/\widehat{G}(s-)$ equals the Kaplan-Meier estimator of $P(C \geq t \mid C > s)$, assuming that no censoring happens at s when $T_i = s$. $\widehat{G}(t-)/\widehat{G}(s-)$ has the form of a usual Kaplan-Meier estimator of a censoring survival function, but taking time s as the new time origin. Hence, the ith individual does not contribute to $\widehat{G}(t-)/\widehat{G}(s-)$ anymore.

Using that the Kaplan-Meier estimator $\widehat{G}$ converges in distribution towards the true censoring survival function and that the expectation of a random variable is the expectation of the conditional expectation of that random variable, we find that, asymptotically, the expectation of the second summand in (5.45) equals

$$G(t-)P(T < t, X_T = 2),$$

which also equals the expectation of the second summand of the representation of $Y^\star_{0;i}(t)$ in (5.43).

Estimation of the regression coefficients from a proportional subdistribution hazards model based on a partial likelihood with 'estimated' risk sets is implemented in the R package cmprsk. The package also provides for an estimator of the covariance matrix of the estimated regression coefficients. However, the covariance estimator is of a more complicated form than $(\mathcal{J}_\tau(\widehat{\beta}))^{-1}$ from Section 5.2.1 on estimation from proportional cause-specific hazards models. This stems from the fact that estimation of the subdistribution risk set also contributes to the covariance. A derivation of the covariance estimator is beyond the technical level of this book. Readers are referred to Fine and Gray (1999), where the estimator is explicitly derived using an empirical process argument (van der Vaart and Wellner, 1996). We also note that Geskus (2011) has recently used martingale arguments in this context.

As illustrated in the data examples below, cmprsk also allows for predicting cumulative incidence functions under a proportional subdistribution hazards assumption in a manner analogous to Section 5.2.1.

Randomly right-censored competing risks data: Multiple imputation of missing censoring times

As explained above and illustrated in the data examples in Section 5.3.3, fitting a proportional subdistribution hazards model is straightforward, if the potential censoring times for individuals with an observed competing event are known. In such a case, standard Cox software as, e.g., provided by the survival package may be used. In the absence of such knowledge, the cmprsk package allows for fitting a proportional subdistribution hazards model.

A drawback of having to use cmprsk is that the package offers less function-ality than survival. E.g., survival also allows users to specify strata with potentially different, strata-specific baseline hazards, include frailties (i.e., ran-dom effects) or maximize a penalized partial likelihood (cf. Therneau and Grambsch (2000)). In Section 11.2.3, we wish to fit a proportional subdistri-bution hazards model with time-dependent covariates. None of these analyses is feasible with cmprsk. Ruan and Gray (2008) therefore suggested a multiple imputation approach which allows for a subdistribution hazards analysis using survival, even if the potential censoring times are not known. The idea is to recover this missing information from an estimator of the censoring survival function $P(C > t)$ in multiple imputation steps.

Briefly, the multiple imputation procedure runs as follows:

1. The aim is to impute censoring times from a given censoring distribution. One computes the Kaplan-Meier estimator $\widehat{G}(t)$ of the censoring survival function $P(C > t)$ using the 'censoring event indicator' $1 - \mathbf{1}(T \leq C)$. This provides for a proper distribution on the observed censoring times, unless the largest event time is *not* a censoring event. In this case, one adds another censoring time to the data, which is larger than the largest event time, before computing $\widehat{G}(t)$.

2. Imputation: Consider all individuals in the data for whom a competing event type 2 has been observed. Let i be such an individual with failure time T_i, $T_i \leq C_i$ and $X_{T_i} = 2$. The potential future censoring time C_i is not known. For individual i, a censoring time is imputed by drawing at random from the conditional distribution $1 - \widehat{P}(C > t \mid C > T_i) = 1 - \widehat{G}(t)/\widehat{G}(T_i)$, $t > T_i$.

3. Multiple imputation: Repeat the previous step k times. This results in k data sets, in which individual i is censored at its respective imputed cen-soring time.

4. Multiple analyses: The analysis, typically running coxph, is performed for each of the k data sets.

5. Pooled estimates: Write $\widehat{\gamma}^{(l)}$ for the estimate of γ of (5.7) obtained from the lth analysis, $l = 1, \ldots k$. Also write $(\mathcal{J}_\tau(\widehat{\gamma}^{(l)}))^{-1}$ for the estimated covariance matrix from the lth analysis (cf. (5.16)). The estimate of the regression coefficients based on multiple imputation then is

$$\widehat{\gamma} = \sum_{l-1}^{k} \widehat{\gamma}^{(l)}/k,$$

and the estimated covariance matrix is (Schafer, 1997)

$$\sum_{l=1}^{k}(\mathcal{J}_\tau(\widehat{\gamma}^{(l)}))^{-1}/k + \frac{k+1}{k}\sum_{l=1}^{k}\left(\widehat{\gamma}^{(l)} - \widehat{\gamma}\right)^{\top}\left(\widehat{\gamma}^{(l)} - \widehat{\gamma}\right)/(k-1),$$

where the first part is the estimated within-imputation variance, and the second part stems from the between-imputation variance.

We illustrate the multiple imputation approach using the R convenience package kmi (Allignol and Beyersmann, 2010) in Section 5.3.3 below.

Ruan and Gray (2008) also considered using an estimator of the censoring survival function based on bootstrap samples in order to better account for the uncertainty in estimating $P(C > t)$. In this approach, the Kaplan-Meier estimator of the censoring survival function is computed from a bootstrap sample drawn with replacement from the original data. A new bootstrap sample is drawn for each imputation step, and the bootstrap sample is only used for estimating $P(C > t)$.

For both real and simulated data, Ruan and Gray found similar results for the simple multiple imputation procedure and the one additionally using the bootstrap. Our examples are in line with this finding. However, we find it hard to give recommendations on whether or when the additional bootstrap step may be necessary. This is particularly true, as the suggestion by Ruan and Gray (2008) is (at the time of writing) still rather new.

Prediction

Predicting cumulative incidence functions under a proportional subdistribution hazards assumption works in a manner analogous to Section 5.2.1. The key step is to obtain a Breslow-type estimator of the cumulative subdistribution baseline hazard $\Lambda_0(t) = \int_0^t \lambda_0(u)\mathrm{d}u$ similar to (5.18). Consider the simple case of administratively right-censored competing risks data first. In this case, all censored subdistribution times are known, and an estimator analogous to (5.18) is given by

$$\widehat{\Lambda}_0(t) := \sum_{\vartheta_i \wedge C_i \leq t} \frac{\Delta N_{01}(\vartheta_i \wedge C_i)}{\sum_{i=1}^n \exp\left(\widehat{\gamma} \cdot Z_i\right) \cdot Y_{0;i}^\star(\vartheta_i \wedge C_i)},$$

where $\widehat{\gamma}$ is the estimated regression coefficient of the proportional subdistribution hazards model.

A predictor of the cumulative incidence function analogous to (5.21) for some (future) individual $\tilde{\imath}$ with covariate $Z_{\tilde{\imath}} = z$ is

$$\widehat{P}(T \leq t, X_T = 1 \mid z) = 1 - \prod_{\vartheta_i \wedge C_i \leq t} \left(1 - \Delta\widehat{\Lambda}_0(\vartheta_i \wedge C_i; z)\right),$$

where $\widehat{\Lambda}_0(t; z) = \widehat{\Lambda}_0(t) \exp(\widehat{\gamma}z)$. Note that the product in the last display needs only be computed over times $\vartheta_i \wedge C_i$ with $\vartheta_i \leq C_i$.

If censoring is not entirely administrative, we may proceed as before. We either substitute $Y_{0;i}^\star(t)$ by (5.44) or we use multiple imputation. In order to approximate the asymptotic distribution of the predicted quantities, one may consider bootstrapping the data; see Appendix A.

5.3.3 Examples

Using the proportional subdistribution hazards model, we reanalyse data which have been analysed earlier in Section 5.2.2 based on the classical cause-specific hazards approach. We provide a guided tour of the different possibilities of how to fit a subdistribution hazards model. We find that all these analyses are in good agreement. The data examples also illustrate the relative merits of the approach based on the subdistribution hazard and based on cause-specific hazards, respectively.

Simulated data

Using standard Cox software for administratively censored data

We reanalyse the simulated data x of Section 5.2.2. Because we have simulated these data, we know both the (uncensored) event time x$T and the censoring time x$C. This situation resembles administrative censoring, where the potential future censoring times are known for all individuals. As a consequence, we may use standard Cox software in order to fit a proportional subdistribution hazards model. As noted earlier, this can be useful, because standard Cox software typically offers more functionality than the specialized R package cmprsk discussed later. The approach is also helpful for appreciating that the proportional subdistribution hazards model really is a model of the proportional hazards type.

In order to use the function coxph of the survival package explained in Section 5.2.2, we need to code the censored subdistribution failure time $\vartheta \wedge C$, cf. Section 5.3.2:

```
> x$thetaandC <- ifelse(x$status==2, x$C, x$TandC)
```

This is a partial output of x, which illustrates the connection between $\vartheta \wedge C$ and $T \wedge C$:

```
   id         T X.T Z          C    TandC status   thetaandC
1   1 8.6883519   2 0 89.504193 8.6883519      2 89.5041930
14 14 0.2355318   1 0 36.702982 0.2355318      1  0.2355318
17 17 7.1147213   2 0  3.929309 3.9293088      0  3.9293088
```

For individual 1, occurrence of event type 2 has been observed. As a consequence, its subdistribution failure time equals infinity, and x$thetaandC equals x$C. The remaining two individuals have an observed type 1 event (individual 14), or no event has been observed for them at all (individual 17). For these individuals, we have that x$thetaandC equals x$TandC.

We now fit model (5.7) using coxph,

```
> summary(coxph(Surv(thetaandC, status == 1) ~ Z, data=x))
```

Note that we only needed to code a different censored event time (i.e., x$thetaandC), but that the status indicator remains unchanged because of (5.42). Of course, the baseline covariate information remains unchanged, too. The output of the call to coxph follows.

```
 n= 1500

      coef exp(coef) se(coef)      z Pr(>|z|)
Z 0.09749   1.10240  0.12353 0.789     0.43

  exp(coef) exp(-coef) lower .95 upper .95
Z     1.102     0.9071    0.8653     1.404
```

As noted earlier, the proportional subdistribution hazards model is misspecified, because the data have been generated from proportional cause-specific hazards models (5.26) and (5.27). We may, however, interpret the estimated subdistribution hazards ratio $\exp(\widehat{\gamma}) = 1.1$ as an estimate of a time-averaged effect seen in the Aalen-Johansen plot of Figure 5.9: the estimate indicates that covariate value 1 appears to have in summary an increasing effect on the cumulative infection probability. The confidence interval $[0.87, 1.4]$ indicates that this effect is not significant based on the available data.

The notion of a 'time-averaged effect' is discussed in more detail in Section 5.4. Here, we only note that the point estimate $\widehat{\gamma}$ is an asymptotically consistent estimate of such an effect, but that the estimated standard error and, hence, the confidence interval rely on the proportional subdistribution hazards assumption. However, our discussion below indicates that these confidence intervals can be quite reasonable. For the present example, we also bootstrapped the data frame x in order to both construct confidence intervals and to estimate the standard error; the results based on 1000 bootstrap samples were virtually identical to those shown above.

Next, we also reconsider the interpretation of the present analysis as showing a time-averaged effect in the analysis using cmprsk below (cf., in particular, Figure 5.13).

Using cmprsk

The proportional subdistribution hazards model can be fitted using the function crr of the R package cmprsk, even if not all potential censoring times are known. The function works with the estimated risk set (5.44). As a consequence, crr is applied to the original data, and not to censored subdistribution failure times as in our use of coxph above:

```
> fit.sh <- crr(ftime = x$TandC, fstatus = x$status,
+               cov1 = x$Z, failcode = 1, cencode = 0)
```

We first consider the arguments supplied to crr before discussing its output: ftime contains the censored event times $T \wedge C$ and fstatus the observed competing event status $1(T \leq C) \cdot X_T$. Baseline covariates are in `cov1`, which will in general be a matrix with columns corresponding to the covariates. The values in `fstatus` indicating the event of interest are given by `failcode=1`, and those values indicating a censored observation are given by `cencode=0`.

The output of `crr` follows.

```
> fit.sh
```

```
convergence:    TRUE
coefficients:
   x$Z1
0.09934
standard errors:
[1]  0.1236
two-sided p-values:
x$Z1
0.42
```

The appearance of the output is obviously different from that of `coxph`, but the results are similar to the previous analysis using `coxph`. Note that we should not expect the analyses to be identical, as `crr` uses an estimated risk set (5.44), which also results in a different asymptotic variance formula. However, our analyses indicate that this difference may not matter much in practice, which is in line with the findings in Fine and Gray (1999). Finally, the reported p-value corresponds to a log-rank-type test for the subdistribution hazard, which we discuss in Chapter 6.

Newer versions (2.2-0 and later) of `cmprsk` conveniently provide approximate 95% confidence intervals using the `summary` function.

```
> summary(fit.sh)
```

```
Competing Risks Regression

Call:
crr(ftime = x$TandC, fstatus = x$status, cov1 = x$Z, failcode = 1,
    cencode = 0)

      coef exp(coef) se(coef)     z p-value
x$Z1 0.0993      1.10    0.124 0.804    0.42

     exp(coef) exp(-coef)  2.5% 97.5%
x$Z1      1.10      0.905 0.867  1.41

Num. cases = 1500
Pseudo Log-likelihood = -1889
Pseudo likelihood ratio test = 0.65  on 1 df,
```

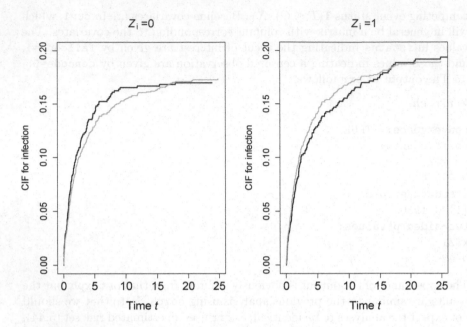

Fig. 5.13. *Simulated data.* Aalen-Johansen estimates of the cumulative incidence functions (CIF) for $X_T = 1$ as in Figure 5.8 (black lines) and predicted CIFs under a proportional subdistribution hazards assumption (grey lines).

cmprsk also allows for predicting cumulative incidence functions. The predicted curves are of a form similar to one minus (5.21), but with the appropriate cumulative subdistribution hazard predictor replacing $\widehat{A}_0.(T_i \wedge C_i; z)$ in (5.21). The predicted curves are obtained using predict.crr,

```
> predict.crr(fit.sh, cov1 = matrix(c(0, 1), nrow = 2))
```

which returns a matrix with unique event times for the event of interest in the first column. The remaining columns are the desired probability estimates corresponding to the covariate values specified in the call to predict.crr. In our example, the second column contains the values of the predicted cumulative incidence function for covariate value 0, and a third column contains entries for covariate value 1.

A customized plot of the results can be found in Figure 5.13, which also shows the Aalen-Johansen estimates from Figure 5.8. The plot suggests that the proportional subdistribution analyses capture the plateaus of the cumulative incidence functions reasonably well, but not how they evolve towards their respective plateaus. This suggests that the time-averaged effect of $\exp(\widehat{\gamma}) = 1.1$ obtained from the proportional subdistribution hazards analysis reflects an increase in the plateaus of the cumulative incidence functions.

Using standard Cox software and multiple imputation

We now fit a proportional subdistribution hazards model using the multiple imputation technique described earlier. The method is implemented in the package kmi.

First, we compute the imputed data sets using the function kmi,

```
> require(kmi)
> imputed.data <- kmi(Surv(TandC, status != 0) ~ 1, data = x,
+                     etype = status, failcode = 1, nimp = 10)
```

Its first argument is a survival object as explained in Section 4.2 and as used in the first-event analysis of the data set x in Section 5.2.2. The survival object distinguishes actually observed event times from censoring events. The argument etype specifies an individual's observed event status $1(T_i \leq C_i)X_{T_i}$. The code for censored observations must be in compliance with the previously specified survival object. failcode indicates the event of interest such that censoring times will be imputed for the other competing events, and nimp specifies the number of imputations.

In the second step, the function cox.kmi fits a proportional subdistribution hazards model for each imputed data set using coxph as illustrated earlier but with the imputed censoring times replacing the originally simulated censoring times. The results are pooled:

```
> fit.impu <- cox.kmi(Surv(TandC, status == 1) ~ Z,
+                     imputed.data)
> summary(fit.impu)

Call:
cox.kmi(formula = Surv(TandC, status == 1) ~ Z,
                    imp.data = imputed.data)


****************
Pooled estimates:
****************
     coef exp(coef) se(coef)     t Pr(>|t|)
Z 0.09868   1.10372  0.12354 0.799    0.424

   exp(coef) exp(-coef) lower .95 upper .95
Z     1.104      0.906     0.8664     1.406
```

The results are very similar to those obtained earlier.

As explained in Section 5.3.2, the multiple imputation approach can be extended to bootstrapping the censoring survival function with the aim to better account for the uncertainty in estimating $P(C > t)$. This is also available within kmi. Specifying bootstrap = TRUE with, say, nboot = 100 bootstrap samples yield results which are again similar to those obtained earlier:

```
> imputed.data.boot <- kmi(Surv(TandC, status != 0) ~ 1,
+                          data = x, etype = status,
+                          failcode = 1, nimp = 10,
+                          bootstrap = TRUE, nboot = 100)
> summary(cox.kmi(Surv(TandC, status == 1) ~ Z,
+                 imputed.data.boot))

Call:
cox.kmi(formula = Surv(TandC, status == 1) ~ Z,
                        imp.data = imputed.data.boot)


****************
Pooled estimates:
****************
    coef exp(coef) se(coef)      t Pr(>|t|)
Z 0.09809   1.10306  0.12354 0.794    0.427

  exp(coef) exp(-coef) lower .95 upper .95
Z    1.103     0.9066    0.8659     1.405
```

Analysis of hospital data: Impact of pneumonia status on admission on intensive care unit mortality

We briefly reanalyse the hospital data from Sections 4.3 and 5.2.2. Recall that an interpretational problem was that pneumonia increased hospital mortality, but showed no effect on the cause-specific hazard for hospital death. The increase in mortality was due to a considerably decreased cause-specific hazard for alive discharge.

As before, we use the data frame my.sir.data. Using the function crr of the package cmprsk,

```
> crr(ftime = my.sir.data$time, fstatus = my.sir.data$to,
+     cov1 = my.sir.data$pneu, failcode = "1", cencode = "cens")
```

we find an estimated subdistribution hazard ratio of 2.65 with 95% confidence interval [1.63, 4.32].

The analysis illustrates both the interpretational appeal and limitations of a subdistribution analysis. On the one hand, the analysis immediately displays that pneumonia increases hospital mortality. But on the other hand, the analysis does not tell how this effect is mediated through the cause-specific hazards.

5.3.4 Proportional subdistribution hazards analysis of all cumulative incidence functions

The fact that the subdistribution analysis allows for a direct probability interpretation as illustrated above has made it popular in applications. This is particularly relevant, because in applied competing risks analyses based on the usual cause-specific hazards an analysis of the competing cause-specific hazard is often missing. (See, e.g., the literature review by Koller et al. (2008) on randomized trials of implantable cardioverter defibrillator implantation and also Section 7.2.) In the hospital data example, a left out analysis of the discharge hazard would have substantially corrupted the interpretation.

A limitation of the subdistribution hazards approach as compared to cause-specific hazards is that only the latter completely determine the stochastic behaviour of the entire competing risks process. The subdistribution hazard only specifies the cumulative incidence function of interest, but not the competing cumulative incidence functions. Both the hospital data example and our in-depth discussion of the simulated data in Section 5.2.2 illustrate that the way cumulative incidence functions evolve over the course of time is best understood from a cause-specific hazards perspective.

The aforementioned facts have led to proportional subdistribution hazards modelling being applied to both cumulative incidence functions in applications. Conceptually, this approach presents problems if one assumes both subdistribution hazards to follow a proportional subdistribution hazards model. Writing $\Lambda_0(t)$ for the cumulative subdistribution baseline hazard for the event of interest with regression coefficient γ of the original model (5.7) and $\check{\Lambda}_0(t)$, $\check{\gamma}$ for the respective quantities attached to the competing cumulative incidence function, Equation (5.6) implies that

$$1 - \exp\left(-\exp(\gamma Z_i)\Lambda_0(\infty)\right) = \exp\left(-\exp(\check{\gamma} Z_i)\check{\Lambda}_0(\infty)\right), \tag{5.47}$$

assuming proportional subdistribution hazards models for both cumulative incidence functions. This implies that the regression coefficient γ of the original model and the limit of all baseline subdistribution hazards, i.e., both $P(X_T = 1)$ and $P(X_T = 2)$ in the baseline situation, determine the regression coefficient $\check{\gamma}$ of the competing model. However, this restriction is not accounted for when running a proportional subdistribution hazards analysis twice, once for the event of interest and once for the competing event. Readers may check that this approach may lead to inconsistent model-based estimates for all cumulative incidence functions by using `predict.crr` also for the competing event in an analysis of simulated data as in Section 5.3.3.

It is therefore natural to interpret such analyses in terms of a time-averaged effect as mentioned above and discussed in more detail in Section 5.4, if proportional subdistribution hazards analyses are carried out for all cumulative incidence functions.

Running `cmprsk` for the competing event in the simulated data of Section 5.3.3 results in a time-averaged subdistribution hazards ratio 0.47 with a model-based 95% confidence interval [0.41, 0.53], reflecting an average decreasing effect seen in Figure 5.8 (right). Readers may check that subdistribution hazards analysis for hospital discharge reflects the decreasing effect of pneumonia seen in Figure 4.9 (right) for the hospital data.

5.3.5 Left-truncation

Estimation from a proportional subdistribution hazards model in the presence of left-truncation has been, at the time of writing, a field of active work in a number of research groups. We briefly discuss the approach of an 'estimated' subdistribution risk set. To the best of our knowledge, Geskus (2011) first suggested extending this approach to left-truncated data.

With potentially also left-truncated data, the subdistribution risk set (5.43) becomes

$$
\begin{aligned}
Y_{0;i}^{*}(t) &= \mathbf{1}(L_i < t \leq T_i \wedge C_i) + \mathbf{1}(L_i < T_i < t \leq C_i, X_{T_i} = 2) \\
&\quad + \mathbf{1}(T_i \leq L_i < t \leq C_i, X_{T_i} = 2) \quad\quad (5.48) \\
&= \mathbf{1}(L_i < t \leq T_i \wedge C_i) + \mathbf{1}(T_i \vee L_i < t \leq C_i, X_{T_i} = 2), \quad (5.49)
\end{aligned}
$$

where we have written $\vee$ for the maximum. The first two terms on the right hand side of (5.48) are analogous to (5.43), the last term accounts for individuals who experience a competing event before study entry.

An 'estimated' subdistribution risk set similar to (5.45) is given by

$$
\begin{aligned}
\widehat{Y^{\star}}_{0;i}(t) &= \mathbf{1}(L_i < t \leq T_i \wedge C_i) \\
&\quad + \mathbf{1}(C_i \geq T_i > L_i)\frac{\widehat{G}(t-)}{\widehat{G}(T_i-)}\frac{\widehat{H}(t-)}{\widehat{H}(T_i-)}\mathbf{1}(T_i < t, X_{T_i} = 2), \quad (5.50)
\end{aligned}
$$

where $\widehat{H}(t)$ is an estimator of $\mathrm{P}(L \leq t)$. Keiding and Gill (1990) derive a Kaplan-Meier-type estimator of H by reversing time: if time runs backwards, a first individual enters the risk set just after the largest censored event time, a second individual enters the risk set just after the second largest event time, and so forth. Note that in this description 'first individual' and 'second individual' refer to reversed time, but 'largest' and 'second largest' refer to real time. A left-truncation event may then happen (in reversed time) after risk set entry following the censored event time.

To justify (5.50), consider, analogous to (5.46),

$$
\mathrm{E}\Big(\mathbf{1}(C_i \geq s > L_i)\frac{\widehat{G}(t-)}{\widehat{G}(s-)}\frac{\widehat{H}(t-)}{\widehat{H}(s-)}\mathbf{1}(s < t, j = 2)\,|\,T_i = s, X_{T_i} = j\Big). \quad (5.51)
$$

Assuming L_i, T_i, and C_i to be independent, one may show, along the route following (5.46), that $\widehat{Y^{\star}}_{0;i}(t)$ is asymptotically unbiased in the sense that its expectation equals $\mathrm{E}(Y^{\star}_{0;i}(t))$ asymptotically. We note that Geskus (2011) discusses that (5.50) remains applicable, if L_i and C_i are dependent.

A representation of $\widehat{Y^\star}_{0;i}(t)$ analogous to (5.44) is

$$
\widehat{Y^\star}_{0;i}(t) = \mathbf{1}(C_i \geq T_i \wedge t > L_i) \cdot
$$
$$
\frac{\widehat{G}(t-)}{\widehat{G}(\{T_i \wedge C_i \wedge t\}-)} \frac{\widehat{H}(t-)}{\widehat{H}(\{T_i \wedge C_i \wedge t\}-)} \cdot \left(\mathbf{1}(t \leq T_i) + \mathbf{1}(T_i < t, X_{T_i} = 2)\right).
$$

$$(5.52)$$

Note that the last term on the right hand side of (5.51) is the subdistribution risk set in the case of complete data as in (5.44). Also note that $\mathbf{1}(C_i \geq T_i \wedge t > L_i)$ equals one, if we have knowledge of individual i's vital status just prior to t.

Estimation now proceeds by replacing $Y^\star_{0;i}(t)$ with $\widehat{Y^\star}_{0;i}(t)$ in the partial likelihood. Deriving a covariance estimator, however, poses a challenge. In their original paper on right-censored data, Fine and Gray (1999) used an empirical process argument. In the absence of such a formula, one may consider bootstrapping the data; see Appendix A. Geskus (2011) uses martingale arguments for a reweighted data set. Once the reweighted data have been prepared, they may be passed to coxph. At the time of writing, an R function to create such a weighted data set has been announced to be included in the mstate package; it has also been available as supplementary material for Geskus' paper at www.biometrics.tibs.org. We also refer readers to Zhang et al. (2011), which appeared during the completion of the book; these authors attack the problem at hand using empirical process arguments.

5.3.6 Simulating proportional subdistribution hazards data

So far, we have used a proportional subdistribution hazards analysis as a synthesis of the effects on the cause-specific hazards. The synthesis was in terms of the effect on one cumulative incidence function. As the analysis of the simulated data illustrated, such a synthesis may even be achieved if the model is misspecified. For the simulated data, we knew that the proportional subdistribution hazards model does not hold, but in a real data analysis there is a priori no reason to assume it to be misspecified. This gives rise to the question of what kind of cause-specific hazards models yield a proportional subdistribution hazards model. Such knowledge may be used for simulation.

In this section, we first present a *direct* way of simulation based on the cause-specific hazards (Beyersmann et al., 2009). We find that some care is required when choosing the cause-specific hazards models. We also present an *indirect* way of simulation, which evades these difficulties, but hides the underlying cause-specific hazards models (Fine and Gray, 1999).

Although we do not apply these simulation algorithms here, we have chosen to present them for readers wishing to further investigate the subdistribution hazards approach by way of simulation.

Direct simulation

The aim is to generate competing risks data such that the subdistribution hazard of the cumulative incidence function of interest (i.e., for event 1) follows model (5.7):

$$\lambda_i(t; Z_i) = \lambda_0(t) \cdot \exp(\gamma \cdot Z_i), \, i = 1, \ldots, n.$$

The algorithm operates in three steps: we first determine the baseline subdistribution hazard and the baseline cause-specific hazards. Next, we determine models for the cause-specific hazards. These are then used to generate the data.

Consider specifying the baseline situation. Recall that we write $\alpha_{01;0}(t)$ and $\alpha_{02;0}(t)$ for the baseline cause-specific hazards, $A_{01;0}(t)$ and $A_{02;0}(t)$ for the cumulative baseline cause-specific hazards and $\Lambda_0(t)$ for the cumulative baseline subdistribution hazard. We have to distinguish between three ways to specify the baseline situation:

1. We choose $\alpha_{01;0}(t)$ and $\alpha_{02;0}(t)$ and compute $\lambda_0(t)$ from Equation (5.39).
2. We choose $\alpha_{01;0}(t)$ and $\lambda_0(t)$. Next, we have to compute $\alpha_{02;0}(t)$. Following from Equation (5.39), we have

$$\lambda_0(t) \exp(-\Lambda_0(t)) = \exp(-A_{01;0}(t) - A_{02;0}(t)) \alpha_{01;0}(t). \qquad (5.53)$$

An easy calculation shows that we may compute $\alpha_{02;0}(t)$ as follows.

$$\alpha_{02;0}(t) = \lambda_0(t) - \alpha_{01;0}(t) - \frac{\mathrm{d}}{\mathrm{d}t} \ln(\lambda_0(t)/\alpha_{01;0}(t)) \qquad (5.54)$$

3. We choose $\alpha_{02;0}(t)$ and $\lambda_0(t)$. Next, we have to compute $\alpha_{01;0}(t)$. Following from Equation (5.53), we have

$$\exp(-A_{01;0}(t)) \alpha_{01;0}(t) = \lambda_0(t) \exp(-\Lambda_0(t) + A_{02;0}(t)). \qquad (5.55)$$

Applying (in this order) integration, ln and d/dt to Equation (5.55) yields

$$\alpha_{01;0}(t) = \frac{\lambda_0(t) \exp(-\Lambda_0(t) + A_{02;0}(t))}{1 - \int_0^t \lambda_0(u) \exp(-\Lambda_0(u) + A_{02;0}(u)) \, \mathrm{d}u}. \qquad (5.56)$$

In the second step, we need to specify models $\alpha_{01;i}(t; Z_i)$ and $\alpha_{02;i}(t; Z_i)$ such that model (5.7) holds. Recall that we write $A_{01;i}(t; Z_i)$ and $A_{02;i}(t; Z_i)$ for the cumulative cause-specific hazards of the ith individual. Analogously, we write $\Lambda_i(t; Z_i)$ for the cumulative subdistribution hazard of the ith individual. We choose a model for one of the two cause-specific hazards. This could be a proportional cause-specific hazards model as in (5.3), but other choices are also possible (cf. Section 5.6).

If we choose a model $\alpha_{01;i}(t; Z_i)$, we determine the model $\alpha_{02;i}(t; Z_i)$ as in Equation (5.54), i.e.,

$$\alpha_{02;i}(t; Z_i) = \lambda_i(t; Z_i) - \alpha_{01;i}(t; Z_i) - \frac{d}{dt} \ln(\lambda_i(t; Z_i)/\alpha_{01;i}(t; Z_i)). \quad (5.57)$$

If, however, we choose a model $\alpha_{02;i}(t; Z_i)$, we determine the model $\alpha_{01;i}(t; Z_i)$ as in Equation (5.56), i.e.,

$$\alpha_{01;i}(t; Z_i) = \frac{\lambda_i(t; Z_i) \exp\left(-\Lambda_i(t; Z_i) + A_{02;i}(t; Z_i)\right)}{1 - \int_0^t \lambda_i(u; Z_i) \exp\left(-\Lambda_i(u; Z_i) + A_{02;i}(u; Z_i)\right) du}. \quad (5.58)$$

In the third and final step, data are generated from the cause-specific hazards models.

In practice, one will usually first determine the baseline cause-specific hazards, which can essentially be any 'well behaving' (differentiable, integrable) non-negative functions, whereas choice of $\lambda_0(t)$ is complicated by the fact that the limit of $1 - \exp\left(-\Lambda_0(t)\right)$ must approach $P(X_T = 1 \mid Z_i = 0)$ as $t \to \infty$. Next, specifying the proportional subdistribution hazards model, a cause-specific hazards model $\alpha_{01;i}(t; Z_i)$ and using Equation (5.57) is the most convenient option. Note that Equations (5.54), (5.56) – (5.58) are subject to the constraint that they result in nonnegative functions.

Indirect simulation

The advantage of the direct simulation approach described above is that it explicitly states the cause-specific hazards leading to a proportional subdistribution hazards model. The practical inconvenience is that typically one cause-specific hazards model will not look 'nice' and that we have to ensure that all hazards are nonnegative. Fine and Gray (1999) circumvented these inconveniences by 'indirect simulation': the cause-specific hazards are only implicitly given. We briefly describe the approach and how the cause-specific hazards may be recovered.

Fine and Gray assumed the cumulative incidence function to follow the model

$$P(T_i \le t, X_{T_i} = 1 \mid Z_i = z) = 1 - \left(1 - p\left(1 - e^{-t}\right)\right)^{\exp(\gamma \cdot z)}, \quad (5.59)$$

where $1 - \left(1 - p\right)^{\exp(\gamma \cdot z)} = P(X_{T_i} = 1 \mid Z_i = z)$ is the probability to experience a type 1 event for individuals with covariate value z, $p \in (0, 1)$. Equation (5.59) is a distribution of a subdistribution failure time with probability mass $1 - P(X_{T_i} = 1 \mid Z_i = z)$ at infinity (cf. (5.35)). The distribution function (5.59) results from a proportional subdistribution hazards model (5.7) with baseline hazard

$$\lambda_0(t) = \frac{pe^{-t}}{1 - p\left(1 - e^{-t}\right)} \quad (5.60)$$

(cf. Equation (5.5)). The competing cumulative incidence function was assumed to be

$$P(T_i \leq t, X_{T_i} = 2 \mid Z_i = z)$$
$$= P(X_{T_i} = 2 \mid Z_i = z) \cdot P(T_i \leq t \mid X_{T_i} = 2, Z_i = z)$$
$$= \left(1 - p\right)^{\exp(\gamma \cdot z)} \cdot \left(1 - \exp\left(-t \cdot \exp(\gamma \cdot z)\right)\right), \tag{5.61}$$

where $P(T_i \leq t \mid X_{T_i} = 2, Z_i = z)$ is an exponential distribution with hazard function $\exp(\gamma \cdot z)$.

On average, $P(X_{T_i} = 1 \mid Z_i = z)$ of the individuals with covariate value z experience the event type 1, whereas $P(X_{T_i} = 2 \mid Z_i = z)$ of these experience event type 2. Conditional on the event type, the associated failure times may then be generated using the conditional distributions $P(T_i \leq t \mid X_{T_i} = j, Z_i = z)$, $j = 1, 2$. Indirect simulation first determines individual i's event type X_{T_i} with $P(X_{T_i} = 1 \mid Z_i = z)$ as given below (5.59). Next, the corresponding event time T_i is generated conditional on X_{T_i} with distribution

$$P(T_i \leq t \mid X_{T_i} = j, Z_i = z) = \frac{P(T_i \leq t, X_{T_i} = j \mid Z_i = z)}{P(X_{T_i} = j \mid Z_i = z)}, j = 1, 2.$$

In principle, this simulation may also be based on the cause-specific hazards. As both cumulative incidence functions and the subdistribution hazard are known, we may use (5.39) to compute $\alpha_{01;i}(t; Z_i)$ and then proceed as described for the direct simulation to compute $\alpha_{02;i}(t; Z_i)$.

5.4 The least false parameter

In the data examples of this chapter, we have fitted several proportional hazards-type models. These models addressed different quantities: all-cause hazard, cause-specific hazards, and subdistribution hazard. Hence, the analyses answered different questions, although connections can be made. In general, the analysis of the cause-specific hazards offered the highest resolution, so to speak, but also required care when interpreting results.

In all of these models, the proportional hazards assumption is made solely for mathematical and interpretational convenience. The mathematical convenience is mirrored in the elegant partial likelihood argument which we have briefly described in Section 5.2.1. Assuming a constant hazard ratio is also interpretationally convenient as it summarizes an effect on the hazard under study in a single number. However, assuming one hazard notion to follow a Cox-type model usually implies that another hazard notion does not.

Properties of statistical procedures, on the other hand, are usually derived assuming that one knows the correct model. The question that we address in this section then is: what is the analysis doing and what should be its interpretation if the underlying model assumption is incorrect? Our impression from the data analyses was that a misspecified proportional hazards analysis (i.e., where the true hazard ratio is time-varying) results in a time-averaged effect. From this point of view, we were able to interpret the analyses of the all-cause hazard and of the subdistribution hazard for the simulated data, which had been generated from proportional cause-specific hazards models, and we could interpret subdistribution hazards analyses for each competing endpoint in the pneumonia data set.

A comprehensive account of misspecified survival data models has been given by Hjort (1992); see, in particular, Sections 6B and 7B. In fact, one finds that the estimator derived from a misspecified model is consistent, although not for the misspecified model parameter, but for the so-called least false parameter. The least false parameter is 'least false' in the sense that it yields the 'best approximation' of the misspecified model towards the true model which generated the data. The approximation is optimal in terms of an appropriate distance between the misspecified and the true model. An easily accessible account of the main ideas is given in Claeskens and Hjort (2008), Sections 2.2 and 3.4.

We do not further investigate these issues in depth, but are content with the fact that our previous interpretation of a misspecified analysis can be given a firm background. See, e.g., Struthers and Kalbfleisch (1986); Lin and Wei (1989); Gerds and Schumacher (2001); Fine (2002); Latouche et al. (2007) for related work. In particular, our approach of looking at models for the cause-specific hazards and for the subdistribution hazard, respectively, side-by-side is investigated by Grambauer et al. (2010a).

We do, however, wish to point out a number of practically relevant issues. First of all, we again emphasize that proportional hazards assumptions are made for convenience. If misspecified, we believe that the summarizing interpretation in terms of the least false parameter is useful. However, this point of view does not dispose of model choice considerations, e.g., if the aim of the analysis is to study time-varying effects; see also Section 5.6.

Second, the least false parameter does also depend on the censoring distribution. Say, if the analysis of the cumulative incidence functions in Figures 5.3 and 5.4 were subject to heavy censoring such that they could only be estimated up to time 5 or 10, our impression gathered from the analysis based on the misspecified model would differ from the one gathered from Figure 5.9. We illustrate this by artificially censoring the simulated data at time 10 such that estimation beyond time 10 is not feasible:

```
> x$TandC.art <- ifelse(x$TandC <= 10, x$TandC, 10)
> x$status.art <- ifelse(x$TandC <= 10, x$status, 0)
```

Originally, only 10.67% of the individuals were censored, but introduction of the artificial censoring time 10 has increased this proportion:

```
> sum(x$status.art == 0) / length(x$id)
```

[1] 0.4813333

The subdistribution hazards analysis

```
> fit.sh.art <- crr(ftime = x$TandC.art, fstatus = x$status.art,
+                   cov1 = x$Z, failcode = 1, cencode = 0)
```

now yields an estimated subdistribution hazards ratio of 0.94 with 95% confidence interval [0.73, 1.22]. The original subdistribution hazards analysis indicated a (non-significant) increase of the cumulative incidence function for event 1. However, the new analysis indicates a (non-significant) decrease. This is in line with Figure 5.9 read up to time 10 only.

For comparison, we also reanalyse the cause-specific hazards. Recall from Chapter 2 that a crucial argument for basing the analysis of event time data on the hazards was that censoring does not 'disturb' the hazard. As the simulated data have been generated from proportional cause-specific hazards models, we should hope for recovering the cause-specific hazard ratios even in the presence of heavier censoring. However, we should also expect a larger variance. Running

```
> summary(coxph(Surv(TandC.art, status.art == 1) ~ Z,data = x))
> summary(coxph(Surv(TandC.art, status.art == 2) ~ Z, data = x))
```

results in estimates of the cause-specific hazards ratio of 0.83 [0.65, 1.08] for event 1 and of 0.18 [0.15, 0.23] for event 2. These results are similar to those obtained earlier, but with somewhat larger confidence intervals for type 2 events. This is so, because type 2 events have also occurred after the artificial censoring time 10, whereas most of the type 1 events have happened before that time.

Third, we already pointed out earlier that a misspecified analysis provides for a consistent estimate of the average effect, but that variance estimates and confidence intervals may be biased. Lin and Wei (1989) and Hjort (1992) provide consistent variance estimates outside model conditions. As an alternative, nonparametric bootstrapping is also considered by Hjort. A robust variance estimator can be computed using `coxph`, setting `robust=TRUE`. Nonparametric bootstrapping for multistate data is addressed in Appendix A.

Robust variance estimation has not been developed for the estimated subdistribution risk set technique implemented in `cmprsk`. However, based on nonparametric bootstrapping, Beyersmann et al. (2009) and Latouche et al. (2011) find that the model-based confidence intervals can be quite accurate.

5.5 Goodness-of-fit methods

The regression models that we have discussed in the preceding part of this chapter have been proportional hazards models, namely for the cause-specific hazards or the subdistribution hazard. In the absence of a competing event, these models reduce to a standard Cox model of the (all-cause) survival hazard. Being hazard models, it should be clear by now that virtually all goodness-of-fit methods for checking the proportionality assumption of an all-cause hazard also apply, when adopted in the obvious manner, to checking the proportionality assumption for one cause-specific hazard, say.

Goodness-of-fit methods are both a vast area and an active research field. They typically consist of graphical checks, more formal hypothesis tests, or a combination of both. Instead of a comprehensive but inevitably incomplete account, we have chosen to exemplarily illustrate that hazard-based techniques generally work in the present context, employing a simple but useful graphical display. Further references are given at the end of this section.

To be more specific, we consider plotting the estimated cumulative hazards within one covariate level against the respective estimate within another covariate level. Recall the cause-specific hazards models (5.24)–(5.27):

$$\alpha_{01}(t; Z_i = 1) = 0.825 \cdot \alpha_{01;0}(t) = 0.825 \cdot \frac{0.09}{t+1},$$
$$\alpha_{02}(t; Z_i = 1) = 0.2 \cdot \alpha_{02;0}(t) = 0.2 \cdot 0.024 \cdot t,$$

$i = 1, \ldots, n$, that we have used for simulating data in Section 5.2.2. Plotting the Nelson-Aalen estimators for the $0 \to 1$ transition, say, for individuals with covariate value 1 (on the y-axis) against the baseline group (i.e., on the x-axis) should approximate a straight line with intercept 0 and slope 0.825.

Interpreting departures from such a straight line is complicated by the fact that the variances of the curves are time-dependent. However, one finds that a convex curve suggests increasing hazard ratios, whereas a concave curve suggests decreasing hazard ratios. In addition, a piecewise linear curve suggests hazard ratios that are piecewise proportional.

We now illustrate this graphical procedure using the simulated data of Section 5.2.2. As these data were simulated from proportional cause-specific hazards models, the respective goodness-of-fit plots should indicate a good fit. In contrast to this, the all-cause hazard does not follow a proportional hazards model anymore. However, we discussed earlier that the cause-specific hazard for event 2 is the major hazard. As this cause-specific hazard follows a Cox model, we do not expect to see a dramatic departure from a straight line. Finally, we investigate the proportional subdistribution hazards assumption for event 1. As illustrated in Figures 5.4 and 5.9, both the true and the estimated cumulative incidence functions cross. Hence, the corresponding subdistribution hazards ratio is not proportional.

In the plots below, we use the usual Nelson-Aalen estimators for the cumulative all-cause hazard and for the cumulative cause-specific hazards, computed within groups defined by the binary covariate. In addition, we also need an estimator $\hat{\Lambda}(t)$ of the cumulative subdistribution hazard $\Lambda(t) = \int_0^t \lambda(u)du$, where we have dropped dependency on the covariate for ease of notation. Such an estimator is easily derived recalling that the subdistribution hazard has been tailor made to mimic a standard survival setting. In other words, the usual Aalen-Johansen estimator (4.18) of the cumulative incidence function should equal one minus a Kaplan-Meier-type estimator from the subdistribution framework:

$$1 - \widehat{P}(T \leq t, X_T = 1) = \prod_{u \leq t} \left(1 - \Delta\widehat{\Lambda}(u)\right)$$

$$= \left(1 - \Delta\widehat{\Lambda}(t)\right) \cdot \left(1 - \widehat{P}(T < t, X_T = 1)\right).$$

As a consequence,

$$\Delta\widehat{\Lambda}(t) = 1 - \frac{1 - \widehat{P}(T \leq t, X_T = 1)}{1 - \widehat{P}(T < t, X_T = 1)}$$

$$= \frac{\widehat{P}(T \geq t)\Delta\widehat{A}_{01}(t)}{1 - \widehat{P}(T < t, X_T = 1)}.$$

Note that the last equation corresponds to relationship (5.39) between $\lambda(t)$ and $\alpha_{01}(t)$. We have

$$\widehat{\Lambda}(t) = \sum_{T_i \wedge C_i \leq t} 1 - \frac{1 - \widehat{P}(T \leq T_i \wedge C_i, X_T = 1)}{1 - \widehat{P}(T < T_i \wedge C_i, X_T = 1)}. \tag{5.62}$$

Estimator $\widehat{\Lambda}(t)$ is easily computed based on the Aalen-Johansen estimator $\widehat{P}(T \leq u, X_T = 1)$, which needs to be evaluated both for times u (for the numerator in (5.62)) and times $u-$ (for the denominator in (5.62)). Note that $\widehat{\Lambda}(t)$ only has nonzero increments at times u, at which a type 1 event has been observed.

Figure 5.14 displays the aforementioned goodness-of-fit plots for the all-cause hazards on the left, and for the cause-specific hazards for both events. As anticipated, the plots seem to indicate a good fit of the proportional hazards assumption for both cause-specific hazards as well as for the all-cause hazards, although the latter model is misspecified. The graphical procedure for the subdistribution hazards of the event of interest is displayed in Figure 5.15. There, we clearly see a departure from the straight line with slope $\widehat{\gamma} = 1.1$. The curve, being convex, suggests that the subdistribution hazard ratio is increasing. Readers may check that this is actually the case, using the cause-specific hazards (5.24)–(5.27) and the relationship (5.39) between subdistribution hazard and cause-specific hazards.

With respect to computation, one has to be careful about plotting the cumulative (subdistribution) hazards at the same time points. For instance, if `cif.etm.z0` and `cif.etm.z1` are the cumulative incidences computed with etm for $Z = 0$ and $Z = 1$, respectively, the `timepoints` option of the `trprob` function is helpful to obtain the CIFs at the same time points.

```
> ## Definition of common time points for both groups
> times <- sort(c(cif.etm.z0$time, cif.etm.z1$time))
```

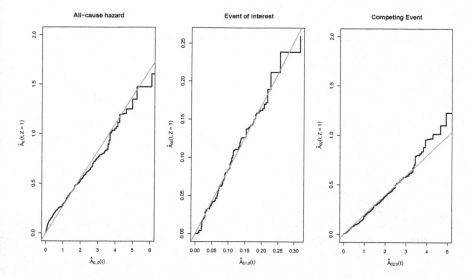

Fig. 5.14. *Simulated data.* Nelson-Aalen estimator of the cumulative all-cause hazard for individuals with covariate value 1 against the baseline group in the left plot, along with the line with slope equal to the corresponding estimated hazard ratio in grey. The middle and right plots are the goodness-of-fit plots for the cause-specific hazards of the event of interest and the competing event, respectively.

```
> ## CIFs at time points 'times'
> cif.z0 <- trprob(cif.etm.z0, tr.choice = "0 1",
+                  timepoints = times)
> cif.z1 <- trprob(cif.etm.z1, tr.choice = "0 1",
+                  timepoints = times)
```

Then, the cumulative subdistribution hazards for the event of interest are easily computed following (5.62).

```
> sub.haz.z0 <- cumsum(1 - ((1 -  cif.z0) /
+                           (1 - c(0, cif.z0[-length(cif.z0)])
+                           )))
> sub.haz.z1 <- cumsum(1 - ((1 -  cif.z1) /
+                           (1 - c(0, cif.z1[-length(cif.z1)])
+                           )))
```

We note that Gill and Schumacher (1987) discuss a relationship between the present graphical check and a more formal test of the proportional hazards assumption. They also consider (optimally) weighted versions of the plotted curves. Very useful textbook accounts of goodness-of-fit methods are, e.g., Section VII.3 of Andersen et al. (1993), Chapter 11 of Klein and Moeschberger (2003), and Chapter 6 of Hosmer et al. (2008). A practically oriented in-depth treatment is provided by Chapters 4–7 of Therneau and Grambsch (2000).

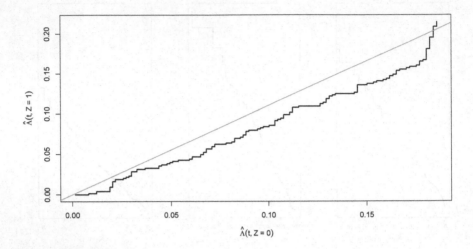

Fig. 5.15. *Simulated data.* Estimated cumulative subdistribution hazards for individuals with covariate value 1 against the baseline group, along with the line with slope equal to the estimated subdistribution hazard ratio in grey.

5.6 Beyond proportional hazards

We have restricted ourselves to discussing Cox-type, proportional hazards models for the cause-specific hazards and for the subdistribution hazard, respectively, because these are the most popular regression models for competing risks data. As stated earlier, these models make a proportional hazards assumption both for interpretational and technical convenience.

Other models exist both for modelling the cause-specific hazards and for direct modelling of the cumulative incidence function. These are of interest per se, as a proportional hazards assumption may or may not fit the data well. If one suspects an effect to be time-varying, or if the aim is to investigate time-varying effects, other models may be attractive. In addition, we found that not all of the popular regression models can fulfill the proportional hazards assumption simultaneously; see also Section 5.4.

An important model, which does not follow a proportional hazards structure, has already been introduced by Aalen in 1980 (Aalen, 1980). For cause-specific hazards, the model postulates

$$\alpha_{0j;i}(t; Z_i) = \beta_{0j0}(t) + \beta_{0j}(t) \cdot Z_i, \ j = 1, 2, \ i = 1, \ldots, n, \tag{5.63}$$

where $\beta_{0j}(t) = (\beta_{0j1}(t), \ldots, \beta_{0jp}(t))$ is a $1 \times p$ vector of regression coefficients and Z_i is a $p \times 1$ vector of covariates for individual i; see also (5.1). The model is entirely nonparametric, because the regression functions are left completely unspecified except for a boundedness assumption. One then proceeds to estimate the *integrated* regression functions. There also exist submodels of (5.63), where some or all of the regression functions are time-constant (Lin and Ying, 1994; McKeague and Sasieni, 1994).

The model is obviously well suited to study time-varying effects. When modelling competing risks in cancer studies, Klein (2006) argues in favour of an additive structure as in (5.63), because both the cause-specific hazards and the all-cause hazard may follow additive models simultaneously.

Still, model (5.63) has somewhat been neglected in applications. In their book on the topic, Scheike and Martinussen argue that this might, in part, be 'due to the fact that the model only contains nonparametric terms [...]' (Martinussen and Scheike, 2006, p. 103). However, there is renewed interest in the model, fueled by Scheike's R package timereg. We refer to Martinussen and Scheike (2006) for an in-depth treatment. We also note that, in addition, these authors consider combinations of proportional hazards models and additive hazards models, that they discuss goodness-of-fit procedures, and that timereg allows us to fit a proportional odds model. If all of the regression functions in (5.63) are time-constant, Schaubel and Wei (2007) show how to use standard statistical software to fit such a model. In their Section 4.2, Aalen et al. (2008) give a very readable account of the additive hazards model, together with arguments in favor of the model. Aalen et al. (2001) consider prediction of transition probabilities both for competing risks and more general multistate models.

It is worthwhile to recall that the additive model can also be applied to the subdistribution hazard using the multiple imputation procedure described in Sections 5.3.2 and 5.3.3. In addition, timereg also allows us to fit models

$$P(T \leq t, X_T = 1 \mid Z_i) = h\left(\gamma_0(t) + \gamma \cdot Z_i\right),$$

with a vector Z_i of covariates for individual i, regression coefficients γ analogous to (5.7), a known *link* function h and a time-dependent regression function $\gamma_0(t)$. The proportional subdistribution hazards model (5.7) is obtained by letting

$$h(x) = 1 - \exp(-\exp(x)).$$

In this case, the time-dependent regression function $\gamma_0(t)$ simply is a transformation of the cumulative baseline subdistribution hazard, i.e., $\exp(\gamma_0(t)) = \Lambda_0(t)$. timereg also allows for other link functions and time-varying effects. The approach is discussed in Scheike et al. (2008); see also Scheike and Zhang (2007, 2008).

As both these references and our discussion of the proportional subdistribution hazards model in Section 5.3 show, there is lively interest in modelling *direct* effects on the cumulative incidence functions. We refer to Andersen and Perme (2008), who give an an overview and further references. They also discuss an important parallel development initiated by Andersen et al. (2003).

These authors suggest a general technique to directly model outcome probabilities based on so-called pseudo-values and using standard generalized linear models software. Similar to the approach by Scheike et al. (2008), this technique is available for more general multistate models (see Scheike and Zhang, 2007). We briefly sketch the ideas for modelling the cumulative incidence function. Klein et al. (2008) present a worked example using the R packages pseudo and geepack. A very useful review paper is Andersen and Perme (2010). Graw et al. (2009) prove some conjectures regarding the asymptotics of the approach for the case of competing risks.

If the aim is to investigate effects on the cumulative incidence function of event 1, pseudo values are obtained by estimating $P(T \leq t, X_T = 1)$ both based on the entire sample and based on a reduced sample, which only leaves the ith individual out. The difference between these two estimates, multiplied by the respective sample size before taking the difference, is the ith pseudo observation, $\widehat{\theta}_i$, say, $i = 1, \ldots, n$. The rationale is that the difference reflects an effect of Z_i on the cumulative incidence function. Andersen et al. then proceed by considering a generalized linear regression model for the cumulative incidence function at time t, potentially over a grid of time points t, using the pseudo values as outcome variables.

We finally encourage readers to occasionally check the CRAN task view 'Survival Analysis' at

http://cran.r-project.org/web/views/Survival.html

The task view, maintained at the time of writing by Aurelien Latouche and Arthur Allignol, aims at presenting R packages for the analysis of time to event data. If methodology such as regression models outside the proportional hazards framework is being made available, either as an R package in its own right or as an addition to an existing R package, the task view is the up-to-date place for finding it.

5.7 Exercises

1. Use the '4D data' simulated in the Exercises of Section 4.6.
   - Run a first-event analysis, i.e., do not distinguish between the competing events, using a proportional hazards model to compare groups.
   - Analyse both cause-specific hazards with a proportional hazards model.
   - Compute the Breslow estimates and predict the cumulative incidence functions using the mstate package,
   - Repeat the analyses with additional left-truncation as in Exercise 5 of Section 4.6.
2. Consider a competing risks process with two competing events and cause-specific hazards $\alpha_{0j}(t)$, $j = 1, 2$. Let $\lambda(t)$ denote the subdistribution hazard. Show that

$$\alpha_{01}(t) = \left(1 + \frac{P(T \le t, X_T = 2)}{P(T > t)}\right) \cdot \lambda(t).$$

3. Fit a proportional subdistribution hazards model for both events in the '4D data' (see above) using both the cmprsk package and the multiple imputation technique.

4. *Simulation following a proportional subdistribution hazards model*: Simulate 200 individuals using the 'indirect simulation' algorithm as in Section 5.3.6 with one binary covariate $Z \in \{0, 1\}$. Covariate values are drawn through a binomial experiment leading to approximately 50% of the individuals having $Z = 1$. We set parameters $p = 0.6$ and $\gamma = 0.3$. Censoring times stem from a uniform distribution with parameters giving about 30% of censored observations.

   Perform analyses of the cause-specific hazards and of the subdistribution hazards. Check graphically that the subdistribution hazards for the event of interest are proportional.

5. The goal of the present exercise is to study how the least false parameter varies depending on the censoring distribution. Generate new censoring times for the data simulated in the Exercises of Section 4.6. Impose heavier censoring $\min(c, C_i)$, where C_i is the censoring time that was originally drawn and c is some small constant. Redo the analyses of the cause-specific hazards and of the subdistribution hazards analyses.

6. Check the goodness-of-fit of Cox models for the all-cause hazard, for the cause-specific hazards and for the subdistribution hazards, respectively, for the data simulated in the Exercises of Section 4.6, using plots as in Figures 5.14 and 5.15.

7. Analyse the impact of transplant type on the cause-specific hazards and the subdistribution hazard of infection for the real ONKO-KISS data okiss, which come with the compeir package. The data set is described in Chapter 1.

6

Nonparametric hypothesis testing

The log-rank test is arguably the most widely used test in survival analysis. In this brief Chapter, we explain the idea of the log-rank test and how it translates to competing risks. The key issue is that the log-rank test compares *hazards* and may consequently be used to compare cause-specific hazards, too. As we have seen earlier, differences between cause-specific hazards do not translate into differences of the cumulative event probabilities in a straightforward manner. Therefore, cumulative incidence functions are often compared by a log-rank-type test for the subdistribution hazard rather than for the cause-specific hazards. As we show below, we may settle for a brief Chapter, because these tests have already been computed as a byproduct of the Cox-type models in Chapter 5.

To begin, assume that there is only one endpoint type, i.e., no competing risks. Say, we wish to compare two groups. Let $N_{0\cdot}^{(1)}(t)$ denote the number of observed events (subject to independent left-truncation and right-censoring) in $[0, t]$ in group 1. Let $N_{0\cdot}^{(2)}(t)$ denote the respective number in group 2. The number of observed events in the pooled sample is $N_{0\cdot}(t) = N_{0\cdot}^{(1)}(t) + N_{0\cdot}^{(2)}(t)$. The respective risk sets are $Y_0^{(1)}(t)$ for group 1, $Y_0^{(2)}(t)$ for group 2 and $Y_0(t)$ for the pooled sample.

In the absence of competing risks, the log-rank test is typically motivated as follows. Consider some t with an observed event (i.e., $\Delta N_{0\cdot}(t) \neq 0$) and also consider the risk sets $Y_0^{(1)}(t)$ and $Y_0^{(2)}(t)$ and the number of observed events $\Delta N_{0\cdot}(t)$ as being given (i.e., as being fixed). Then the expected number of events in group 1 under the null hypothesis of no difference between the two groups is

$$\mathrm{E}\left(\Delta N_{0\cdot}^{(1)}(t)\right) = Y_0^{(1)}(t) \cdot \frac{\Delta N_{0\cdot}(t)}{Y_0(t)}.$$

Hence, under the null hypothesis, we would expect that

$$\Delta N_{0\cdot}^{(1)}(t) - Y_0^{(1)}(t) \cdot \frac{\Delta N_{0\cdot}(t)}{Y_0(t)} \approx 0. \tag{6.1}$$

The log-rank statistic then is

$$\frac{1}{\sum_t V_t} \cdot \left(\sum_t \Delta N_{0\cdot}^{(1)}(t) - Y_0^{(1)}(t) \cdot \frac{\Delta N_{0\cdot}(t)}{Y_0(t)} \right)^2,$$

where summation is over all observed event times in the pooled sample, and V_t denotes the variance obtained from the hypergeometric distribution,

$$V_t = \frac{\Delta N_{0\cdot}(t)(Y_0(t) - \Delta N_{0\cdot}(t))Y_0^{(1)}(t)Y_0^{(2)}(t)}{Y_0(t)^2(Y_0(t) - 1)}.$$

Under the null hypothesis, the log-rank test statistic is approximately χ^2 distributed with one degree of freedom.

For our purposes, it is helpful to rewrite (6.1) as

$$\frac{\Delta N_0^{(1)}(t)}{Y_0^{(1)}(t)} \approx \frac{\Delta N_{0\cdot}(t)}{Y_0(t)}. \tag{6.2}$$

Statement (6.2) corresponds to a counting process point of view, comparing the increment of the Nelson-Aalen estimator based on the data in group 1, i.e., $\Delta N_0^{(1)}(t)/Y_0^{(1)}(t)$ with the respective increment based on the pooled sample, i.e., $\Delta N_{0\cdot}(t)/Y_0(t)$. Summation over all observed event times then implies that the Nelson-Aalen estimators rather than their increments are being compared. This point of view easily adapts to competing risks. Instead of comparing the all-cause Nelson-Aalen estimators, we simply compare the cause-specific Nelson-Aalen estimators for type j events by replacing $N_0^{(1)}(t)$ with $N_{0j}^{(1)}(t)$ and $N_{0\cdot}(t)$ with $N_{0j}(t)$, $j = 1, 2$. It turns out that the respective log-rank test corresponds to the score test obtained from fitting a Cox model to the cause-specific hazard for event type j with group membership as the sole covariate. This fact is more than mere algebraic coincidence, but is related to the log-rank test being locally most powerful for departures from the null hypothesis that follow proportional hazards. In addition, we easily obtain k-sample tests, $k \geq 2$, via this route: one needs to choose one group, say, group 1, as the 'baseline group'. Then, fitting a Cox model with $k - 1$ covariates indicating group membership for group $2, 3, \ldots, k$ will yield the appropriate test.

Readers are encouraged to recheck the output of `summary` applied to `coxph` in Section 5.2.2, which reported the log-rank test at the bottom of its output. Alternatively, one may also use the function `survdiff` of the package `survival` in order to obtain the log-rank test. We briefly revisit the hospital data example of that section, where the aim was to investigate the impact of pneumonia diagnosed on admission. The result was no significant difference between the cause-specific hazards for hospital death, but significantly different cause-specific hazards for alive discharge. This example, like the other analyses in Section 5.2.2, also illustrates the limitations of testing equality of cause-specific hazards. Our previous analysis showed that pneumonia did increase the number of patients dying in hospital, although the hazards for hospital death had been found to be similar with or without pneumonia. A key finding in Section 5.2.2 has been that an adequate interpretation of a competing risks analysis requires careful consideration of *all* cause-specific hazards, including the signs and magnitudes of the effects of, e.g., pneumonia status as well as the relative magnitude of the single cause-specific hazards within one group. Typically, this somewhat complex picture is only partially reflected in a formal hypothesis test of one or even all cause-specific hazards.

In Section 5.3, we have discussed that this complex situation has motivated efforts to study *directly* the impact of covariates such as pneumonia in the hospital data example on the cumulative incidence functions; see also Sections 5.1 and 5.6. In particular, the proportional subdistribution hazards model of Section 5.3 is often used to this end. Recall that this model is a Cox-type model for the subdistribution hazard, and that the subdistribution hazard reestablishes a one-to-one correspondence with the cumulative incidence function of interest. Hence, the score test obtained from fitting such a model must be suited to compare cumulative incidence functions *directly*. We briefly illustrate this with the hospital data. Running

```
> crr(ftime = my.sir.data$time, fstatus = my.sir.data$to,
+       cov1 = my.sir.data$pneu, failcode = "1", cencode = "cens")

convergence:  TRUE
coefficients:
my.sir.data$pneu1
          0.9749
standard errors:
[1] 0.2496
two-sided p-values:
my.sir.data$pneu1
          9.4e-05
```

we find a significant effect of pneumonia on the cumulative incidence function for hospital death. This test is due to Gray (1988) and therefore often called Gray's test. The test may be derived in analogy to our derivation of the standard log-rank test above, if we use the increments of the estimator of the cumulative subdistribution hazard as in Section 5.5. However, similar to the proportional subdistribution hazards model, the challenge is to calculate the asymptotic distribution of the test statistic (Gray, 1988). We also note that the function `cuminc` of the package `cmprsk` may alternatively be used in order to obtain Gray's test.

In Sections 5.3.4 and 5.4, we discussed that proportional subdistribution hazards models typically do not hold for all cumulative incidence functions, nor do Cox-type models typically hold for both cause-specific hazards and subdistribution hazard. Model misspecifications are therefore inevitable when running these analyses side-by-side as is common practice. We discussed the interpretational consequences in Section 5.4. In the present context, running analyses side-by-side means using the usual log-rank test to compare the single cause-specific hazards and maybe also the all-cause hazard as well as using Gray's test to compare the cumulative incidence functions. Here, model misspecification is not a concern in the sense that these tests are entirely non-parametric and valid for their respective null hypotheses. As stated above, log rank-type tests are well suited to detect departures from the null hypothesis that follow proportional hazards *of the respective hazard notion*. Conversely, this entails that a log-rank test for the cause-specific hazard of event 1 will often fail to detect a difference if the hazards cross. Analogous statements hold for the log-rank test of the all-cause hazard and of the subdistribution hazard, respectively. To illustrate these restrictions, recall the first example in Section 5.2.2, where the cause-specific hazards followed perfect Cox models, but the cumulative incidence functions for the event of interest crossed.

Finally, we refer to Hosmer et al. (2008, Section 2.4) for a practical overview on alternatives to the log-rank test and to Bajorunaite and Klein (2008) who give an excellent overview of the available tests for comparing cumulative incidence functions. Bajorunaite and Klein also discuss tests for crossing subdistribution hazards, and they make the distinction between comparing cause-specific hazards and comparing cumulative incidence functions very clear.

6.1 Exercises

1. Use the '4D data' simulated in the Exercises of Section 4.6.
   - Compute the log-rank test for both cause-specific hazards.
   - Compute Gray's test for both cumulative incidence functions.

7

Further topics in competing risks

7.1 More than two competing risks

So far, our treatment of competing risks has been restricted to two competing event states only. An exception was the analysis of drug-exposed pregnancies in Sections 4.4 and 5.2.2, where we handled three competing event states without much ado. This effortlessness shows that otherwise focusing on two competing events has been for ease of presentation only. The aim of the present section is to briefly demonstrate that everything that has been said before for two competing risks easily generalizes to J competing risks, where J is some finite number as in Section 2.2.3.

First of all, the basic competing risks multistate model of Figure 3.1 generalizes to Figure 2.5 by introducing as many absorbing states as there are competing risks under study.

The corresponding competing risks process $(X_t)_{t \geq 0}$ now has state space $\{0, 1, \ldots, J\}$. As before, the process starts in state 0 at time origin, leaves the initial state 0 at time T, and enters the competing event state $X_T \in \{1, \ldots, J\}$. The definitions (3.5) and (3.6) of the cause-specific hazards $\alpha_{0j}(t)$ and their cumulative counterparts $A_{0j}(t)$ remain unchanged, but j now takes values in $\{1, \ldots, J\}$ instead of $\{1, 2\}$. The (cumulative) all-cause hazard is the sum over all (cumulative) cause-specific hazards. The definitions (3.10) and (3.11) of the survival function and of the cumulative incidence functions remain unchanged, too. Of course, there is now one cumulative incidence function per competing event.

The simulation algorithm of Section 3.2 works analogously. The only difference is that the binomial experiment in step 3 of the algorithm has to be replaced by a multinomial experiment. The multinomial experiment decides with probability

$$\alpha_{0j}(T)/(\alpha_{01}(T) + \ldots + \alpha_{0J}(T)) = \alpha_{0j}(T)/\alpha_{0.}(T)$$

on event type $X_T = j$, $j \in \{1, 2, \ldots, J\}$, at event time T.

Nonparametric estimation of Chapter 4 works analogously, too, by introducing cause-specific counting processes for all event types,

$$N_{0j;i}(t) := \mathbf{1}(T_i \wedge C_i \leq t, L_i < T_i \leq C_i, X_{T_i} = 1), j = 1, \ldots, J,$$

such that $N_{0j;i}(t)$ counts the number of observed type j events for individual i in $[0, t]$. No modification is required for the ith at-risk process (4.1), as it only keeps track of whether an event has been observed at all, but not of its type.

Aggregation over all individuals is done in the obvious way. The Nelson-Aalen estimators of the cumulative cause-specific hazards are as in (4.8) and the Aalen-Johansen estimators of the cumulative incidence functions are as in (4.18), but j now takes values in $\{1, \ldots, J\}$. The same remark holds for variance estimation and construction of confidence intervals. All-cause hazards and all-cause probabilities are, of course, sums over the respective J cause-specific quantities.

The necessary adaptions in Chapter 5 on proportional hazards modelling are done in the obvious way.

The bottom line is that we may single out a certain competing risk $\tilde{j}$, $\tilde{j} \in \{1, \ldots, J\}$ and group the remaining $J - 1$ competing risks into one combined competing event state. We may then work with two competing risks as described before, which yields the right results for event type $\tilde{j}$. Doing this for every competing event type $\tilde{j} \in \{1, \ldots, J\}$ in turn yields the adaptions described in this section.

7.2 Frequently asked questions

This section addresses issues that, in our experience, are frequently raised in the context of competing risks. However, we often did not find them helpful in answering the research question of the subject matter. The multistate perspective is well suited to settle these issues.

For ease of presentation, we again consider two competing risks only. I.e., at event time T, either event type $X_T = 1$ or event type $X_T = 2$ occurs. Recall from Section 7.1 that this set-up easily generalizes to more than two competing events.

Are competing risks independent?

The question of statistical independence of the competing risks is almost notorious, but without meaning from a process perspective. The different failure causes are simply different values of exactly one random variable, X_T. Hence, the concept of statistical independence does not apply. The question is only meaningful if one assumes that the competing risks data (T, X_T) arise from risk-specific latent failure times, such that T is the minimum of these hypothetical times and X_T is arg min (cf. Section 3.3). This model has been criticized, because the dependence structure of the latent times is not identifiable from the observable data (Prentice et al., 1978), and also because it has been found to lack plausibility in biometrical settings. We also think (this is connected to the plausibility concern) that the latent failure time model does not pass Occam's razor. Russell (2004, p. 435) summarizes the principle: '[Occam said:] 'It is vain to do with more what can be done with fewer.' That is to say, if everything in some science can be interpreted without assuming this or that hypothetical entity, there is no ground for assuming it.' Russell comments: 'I have myself found this a most fruitful principle in logical analysis.' For competing risks, Occam's razor implies that latent failure time modelling should only be used, if interest actually lies in the latent failure times. E.g., in medical applications, assuming latent times is in our experience usually not needed to answer the research question. See Section 4.3.1 for a real data analysis which illustrates that assuming latent times does not provide further insight. The interpretation of the data analysis, however, turns out to be less straightforward. In addition, the latent failure time approach has to cope with an awkward interpretation of the latent times and has to make an unverifiable assumption on their statistical dependence.

Can we treat a competing event as censoring?

This question arises both in the context of the latent failure time models discussed above and in, e.g, proportional cause-specific hazards analyses as in Section 5.2.2.

If we assume competing risks to arise from two independent latent failure times, the situation resembles the standard random censorship model. Assuming an event time T and a censoring time C to be independent, we may use the Kaplan-Meier estimator to estimate the survival function of T. Analogously, we might treat type 2 events as censoring and use the Kaplan-Meier estimator to estimate the survival function of the latent event time for type 1 events. As discussed above, care must be displayed as to whether assuming latent times is a fruitful model at all.

In Section 5.2.2, we illustrated that this debate may be safely ignored, but that, say, cause-specific hazards may be analysed by fitting two separate Cox models, if all covariates display separate cause-specific effects. In the R code, the analysis of $\alpha_{01}(t)$, say, only treated type 1 events as events and handled competing events and the usual censorings alike. The justification for this is a partial likelihood argument (cf. (5.17)) and has nothing whatsoever to do with assuming independent latent failure times.

In fact, we saw in Section 2.2.3 that occurrence of events other than type 1, say, acts as independent (but not random) right-censoring with respect to estimating the cumulative cause-specific hazard $A_{01}(t)$ for type 1. This means that removal of prior non-type 1 events from the at-risk set allows for estimation of $A_{01}(t)$. In other words, occurrence of non-type 1 events does not disturb the martingale estimation equation (2.24). However, it is important to note that occurrence of competing events is also informative in the sense that probability estimates depend on all hazards. A rigorous discussion of these issues can be found in Andersen et al. (1993), Example III.2.6 and Section III.2.3.

The bottom line is that we may analyse one cause-specific hazard by coding other competing events as censoring, but that the analysis will remain incomplete until this has been done for every competing risk in turn. This has, in particular, been illustrated by the analysis of the simulated data in Section 5.2.2. We were able to make valid analyses for the cause-specific hazards by treating other competing events as censoring, but the analysis would have been incomplete and even misleading in terms of the cumulative event probabilities, until all cause-specific hazards had been analysed.

Can we use one minus Kaplan-Meier to estimate the cumulative incidence function?

The question is, among other things, triggered by the fact that competing events can be coded as censoring for analyses of the cause-specific hazards, but the answer is: no. Standard arguments often tend to be somewhat technical, but the bottom line is rather simple: one minus Kaplan-Meier aims at estimating a distribution function, which eventually approaches 1. In contrast, $P(T \leq t, X_T = j)$ is bounded from above by $P(X_T = j)$, which, in a true competing risks setting, is less than 1, $j = 1, 2$. As a consequence, one minus Kaplan-Meier overestimates the cumulative incidence functions. Often, these biased estimates will eventually add up to more than 1, but $P(X_T = 1) + P(X_T = 2) = 1$. In contrast, our derivation of the Aalen-Johansen estimates in Section 4.1 shows that their sum always equals one minus the meaningful Kaplan-Meier estimate (4.13) of $P(T > t)$.

Can we model censoring as a competing event?

This question has connections to the first three questions. The answer is: yes, but usually one is not interested. The idea is to introduce an additional competing event state in the standard competing risks multistate model of Figure 3.1, to which an individual moves, if a usual right-censoring event occurs, i.e., $C < T$. In most of the cases, one would not be interested in such a model, and one would rather treat censoring as a nuisance parameter. This means that typically the interest is to estimate the distribution of T or (T, X_T) but not C.

Of note, the interpretation of the states in the extended model changes. E.g., being in the initial state would now be interpreted as 'no competing or censoring event observed yet'. The waiting time in this new initial state would now be $T \wedge C$, such that the data are complete in this model in the absence of left-truncation.

An important exception where one is, in fact, interested in the censoring mechanism itself is inverse probability of censoring weighting. Here, the idea is to use weights that compensate for censored observations. The technical difficulties posed by censoring are tackled by studying the censoring distribution directly. In this book, this approach has been used to estimate subdistribution risk sets in Section 5.3.2. Aalen et al. (2008) give a concise and readable textbook account of the inverse probability of censoring weighting principle; see their Section 4.2.7. We note that interest in the censoring distribution is typically limited to the extent that it is used as a tool. The estimated subdistribution risk sets in Section 5.3.2 were used to make inference for $P(T \leq t, X_T = 1)$, but not for $P(C \leq t)$, say.

Don't competing risks apply only if the failure causes are fatal or terminal?

This question, which has a metaphysical flavour, stems from the fact that different causes of death are a standard competing risks example. The multistate perspective offers a simple answer: competing risks model time T until some first event and the type X_T of that first event, but competing risks do not model subsequent events. In order to do this, more complex multistate models are needed.

Can individuals who have failed from a competing event in the past still fail from the event of interest in the subdistribution model?

This question is motivated by the subdistribution at-risk set (5.41) which still includes an individual who has experienced event 2 in the past until the individual's potential future censoring time (cf. the discussion following (5.41)). However, such an individual does *not* move into event state 1 at any time, as is apparent from our definition (5.32) of the subdistribution process. Such an individual simply stays in the initial state of the subdistribution process and never makes a transition into state 1. As it stays in state 0, it still contributes to the risk set with the sole aim to weight the subdistribution hazard down as compared to the cause-specific hazard of interest. As explained in the context of Equation (5.6), the aim of this weighting is to establish the desired one-to-one correspondence between subdistribution hazard and cumulative incidence function.

Can we use the log-rank test in the presence of competing risks? And if so, is it appropriate to compare cumulative incidence functions?

The log-rank test works for data subject to independent right-censoring/left-truncation as explained in Section 2.2.2. Hence, the log-rank test can be applied to test equality of the cause-specific hazard for event type 1, say, between groups, treating type 2 events as censored (cf. our discussion above on ' Can we treat a competing event as censoring?'). It is a subtle issue to which extent such a test is of interest. In general, it will not allow for a probability interpretation. Such a test does not compare cumulative incidence functions. An exception is the special situation where we are willing to assume that the competing cause-specific hazard is the same for all groups (cf. Section 4.5). This situation has attracted some interest in sample size planning (Schulgen et al., 2005; Latouche and Porcher, 2007). There are tests that do compare cumulative incidence functions (Gray, 1988; Pepe and Mori, 1993; Lin, 1997). These tests address a different testing problem; in particular, the null hypothesis is different from the standard log-rank test. See also Chapter 6.

7.3 Discrete time methods for competing risks

We illustrate the use of discrete time methods with incomplete data using the hospital data set that has already been analysed in Sections 4.3 and 5.2.2. Interest focuses on the effect of pneumonia on intensive care unit mortality, the competing risk being discharge alive from the unit.

D'Agostino et al. (1990) illustrated that the so-called pooled logistic regression, in which repeated observations for an individual are pooled and then analysed using a logistic regression model, and the Cox proportional hazards model may give similar results in the standard survival setting. D'Agostino et al. also gave conditions under which the two approaches are asymptotically equivalent. One typical requirement is that the time between measurements is short. This requirement is fulfilled in hospital epidemiology, where data are usually recorded on a day-to-day basis.

For competing risks, the same technique may be adapted using a multinomial logit model instead of logistic regression. Barnett et al. (2009) give a detailed and practical account for the hospital epidemiology setting. Patients' possible outcomes for each day are 'stayed', 'discharged', and 'died'. Assuming that this nominal response follows a multinomial distribution, probabilities of 'stayed', 'discharged' and 'died' can be estimated for each patient day, and the obtained regression coefficients should be close to those obtained when fitting Cox models for the cause-specific hazards.

As a further refinement, one could include a dummy 'day' variable for the first 20 days in the intensive care unit, say (and possibly more, depending on the size of the data set). This lets the baseline risk, which otherwise is assumed to be constant, vary with time, thus mimicking a piecewise constant baseline hazard in a survival model.

The first step for fitting the multinomial logit model is to expand the original data set in order to obtain a data set that, for each patient, has one row per day spent in the unit. We also include the dummy 'day' variable, which is simply a count of the number of days in unit.

```
> ### creation of the dummy variable
> dummies <- c(1:20, rep(20, max(my.sir.data$time) - 20))
> ### creation of the extended data set.
> sir.data.ext <- by(my.sir.data, my.sir.data$id, function(x) {
+     temp <- x[rep(1, x$time), ]
+     temp$day <- dummies[1:nrow(temp)]
```

```
+         status <- c(rep("cens", nrow(temp) - 1),
+                     as.character(temp$to[1]))
+         temp$status <- factor(ifelse(status == "cens", 0, status),
+                               levels = c(0, 1, 2))
+         temp
+       })
> sir.data.ext <- do.call(rbind, sir.data.ext)
> head(sir.data.ext, 4L)

       id from to time pneu day status
41.1   41   0  2    4    0   1      0
41.1.1 41   0  2    4    0   2      0
41.1.2 41   0  2    4    0   3      0
41.1.3 41   0  2    4    0   4      2
```

Above is an excerpt of the extended data set for individual 1. The individual stayed four days in the unit. Hence, the newly created **status** variable is 0 for the first three days, then switches to 2 at day 4, because the individual is discharged alive at this time. The **pneu** variable is always 0, which encodes pneumonia status at baseline.

We now fit the multinomial logit model using the **multinom** function included in the **nnet** package (Venables and Ripley, 2002). We also include the dummy variable **day** as a factor in the model.

```
> require(nnet)
> fit.multinom <- multinom(status ~ pneu + factor(day),
+                          sir.data.ext)

# weights:  66 (42 variable)
initial  value 11916.647495
iter  10 value 2937.595559
iter  20 value 2841.508670
iter  30 value 2801.509575
iter  40 value 2783.487669
iter  50 value 2774.556102
iter  60 value 2774.465681
final  value 2774.455505
converged

> coef(fit.multinom)[, "pneu"]

          1          2
-0.1563889 -1.1403495
```

Computing **exp()** of the coefficients reported above, we obtain an odds ratio of 0.86 for the outcome 'death'. Analogously, using the output of **confint(fit.multinom)**, we get an 95% confidence interval of [0.51, 1.44].

These results are in close agreement with the Cox analyses of the cause-specific hazards in Section 5.2.2. The respective result for the outcome 'discharged' is an odds ratio of 0.32 with 95% confidence interval [0.25, 0.41]. There is again close agreement with the Cox analyses of the cause-specific hazards.

Part III

Multistate models

8

Multistate models and their connection to competing risks

Except for Section 2.2.4, this book has so far focused on competing risks, which model time until first event and type of that first event. We now turn to more complex multistate models, which, e.g., would also allow us to study the occurrence of subsequent events. Such investigations are often of subject matter interest. E.g., in medical applications they will allow for a more detailed study of the course of a patient's disease. In the ONKO-KISS example of bloodstream infection in stem-cell transplanted patients studied in Section 5.2.2, such a model would allow us also to study mortality after infectious complication. In the SIR3 example of Sections 4.3, 5.2.2, 5.3.3 and Chapter 6, we investigated the impact of pneumonia admission diagnosis on intensive care unit mortality. A more complex multistate model would allow the further study of subsequent events which happen during unit stay such as ventilation switched on or off, catheter usage, or occurrence of hospital-acquired infections.

A basic theme of the multistate part of the book is that many parts from the toolbox of competing risks methodology can also be used to analyse multistate models. We therefore build on the competing risks material, which has been discussed in detail earlier in this book, as often as possible. To this end, it is important to understand the connection between competing risks and the type of multistate models at hand. The most common multistate models are time-inhomogeneous Markov processes with a finite state space. The Markov property means that the future state of an individual at time t only depends on the current time s, $s \leq t$, and on the individual's current state, say, 'healthy' or 'ill', but, e.g., not on how long a diseased individual has been ill. We give a more precise statement of the Markov property below. A key issue is that Markovian multistate models can be thought of as being realized as a nested series of competing risks experiments. This connection can be used to analyse the transition hazards of such a multistate model using competing risks-type techniques.

In the following, we usually simply write 'multistate model' when we deal with a Markovian multistate model, dropping the attribute 'Markov'. We do, however, explicitly state when we consider relaxing the Markov assumption as in Chapter 12.

8.1 Multistate models: Time-inhomogeneous Markov processes with finite state space

Instead of *competing risks* processes, we now consider *multistate* processes $(X_t)_{t\geq 0}$, where X_t denotes the state an individual is in at time t. The state space is $\{0, 1, 2, \ldots, J\}$ (i.e., $X_t \in \{0, 1, 2, \ldots, J\}$). This is not unlike the competing risks process with J competing event states in Section 7.1. The difference is that there need not be a common initial state, which is occupied by all individuals at time origin, and, conceptually, any transition $l \to j$, $l \neq j$, $l, j \in \{0, 1, 2, \ldots, J\}$ could be modelled. States out of which transitions are modelled are called *transient states*. In contrast, *absorbing states* are states out of which transitions are not modelled.

Figure 8.1 displays an important example of a multistate model, the illness-death model without recovery. In Figure 8.1 (left), an individual may either

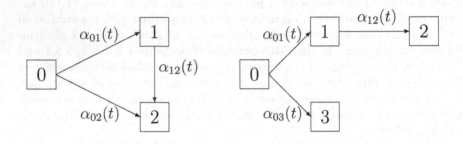

Fig. 8.1. Illness-death model without recovery.

start in state 0 or in state 1, i.e., $X_0 \in \{0, 1\}$. The name 'illness-death model' reflects that, often, being in state 0 is interpreted as being 'healthy', individuals who are 'ill' are in state 1, and state 2 represents 'death'. If all individuals are 'healthy' at time 0, we have again a common initial state. *Events* are modelled as transitions between the states of the model. Occurrence of disease is modelled as a $0 \to 1$ transition, occurrence of 'death' is modelled by transitions into state 2. In the figure, states are represented by boxes and possible transitions are represented by arrows.

Figure 8.1 (right) gives a different representation of essentially the same model. States 0 and 1 are again interpreted as 'healthy' and 'ill', respectively, but there are now two 'death' states 2 and 3. State 2 can only be reached by individuals who are 'ill', state 3 is only accessible for 'healthy' individuals. It is important to note that the models in Figure 8.1 are equivalent. For competing risks processes, we found that their stochastic behaviour is determined by the cause-specific transition hazards, which could be thought of as forces moving along arrows as in Figure 8.1. This also holds analogously for the transition hazards of a multistate model which we define below. For the time being, note that the number of arrows and the possible transitions are the same in both representations of the illness-death model. The only difference is that the data coding following the model of Figure 8.1 (right) conveniently encodes whether a 'dead' individual has been 'healthy' or 'ill' just before 'dying'. This representation is sometimes referred to as the *progressive* illness-death model (e.g., Hougaard, 2000).

There are two important extensions of the multistate model of Figure 8.1. An illness-death model *with* recovery also models 'ill' → 'healthy' transitions by also including an arrow 1 → 0. And we may also model competing end-points by including an additional absorbing state in Figure 8.1 (left). In Figure 8.1 (right), modelling two competing endpoints would amount to including four absorbing states in the model. We later consider these extensions, e.g., Figure 9.5 displays a progressive model with competing endpoints. But, for the time being, we mainly use the basic illness-death model as an exemplary multistate model. Even more complex models can be obtained by including additional states (boxes in the figure) and transitions (arrows in the figure).

In order to establish the connection between competing risks and multi-state models, we first define the transition probabilities and the transition hazards. The matrix of transition probabilities of the process $(X_t)_{t \geq 0}$ with state space $\{0, 1, 2, \ldots, J\}$ is

$$\mathbf{P}(s, t) := (\mathrm{P}_{lj}(s, t))_{l,j}, \, l, j \in \{0, 1, 2, \ldots, J\}, \tag{8.1}$$

with transition probabilities

$$\mathrm{P}_{lj}(s, t) := \mathrm{P}(X_t = j \mid X_s = l, \mathrm{Past}), s \leq t, \tag{8.2}$$

where we have written 'Past' for knowledge about the process' history up to time s. More formally, the 'Past' is a σ-algebra generated by the process, but we stick to the informal 'Past'.

As stated above, we typically assume $(X_t)_{t \geq 0}$ to be Markov, which means that

$$\mathrm{P}(X_t = j \mid X_s = l, \mathrm{Past}) = \mathrm{P}(X_t = j \mid X_s = l), l, j \in \{0, 1, 2, \ldots, J\}, s \leq t. \tag{8.3}$$

In words, the Markov property (8.3) states that the transition probabilities only depend on the past via the current time s and the currently occupied state.

The transition hazards of the multistate model are defined as

$$\alpha_{lj}(t) \cdot \mathrm{d}t := \mathrm{P}(X_{(t+\mathrm{d}t)-} = j \mid X_{t-} = l), l, j \in \{0, 1, 2, \ldots, J\}, l \neq j, \tag{8.4}$$

and the cumulative transition hazards are

$$A_{lj}(t) = \int_0^t \alpha_{lj}(u)\,\mathrm{d}u. \tag{8.5}$$

Note again that the Markov property implies that conditioning on $X_{t-} = l$ is tantamount to conditioning on the entire past of the process before t. Also note that the transition hazards do depend on time t, which means that the process is *time-inhomogeneous*. In contrast to this, a homogeneous Markov process has time-constant transition hazards. In the literature, a homogeneous Markov process is sometimes simply called 'Markov process', dropping the attribute 'homogeneous'.

Before we eventually establish the connection between multistate models and competing risks, already note that the cumulative incidence function of, say, type 1 events in the competing risks multistate model in Figure 3.1 is, in fact, a transition probability, namely $P_{01}(0, t)$. Even more striking is the similarity between the cause-specific hazards (3.5) in competing risks and the transition hazards (2.28). Multiplied with $\mathrm{d}t$, the *cause-specific* hazard $\alpha_{0j}(t)$ is the probability of making a $0 \to j$ transition in the very small time interval $\mathrm{d}t$, $j \neq 0$. Multiplied with $\mathrm{d}t$, the *transition* hazard $\alpha_{lj}(t)$ is the probability of making an $l \to j$ transition in the very small time interval $\mathrm{d}t$, $l \neq j$. Intuitively, $\mathrm{d}t$ will be so small that the transition occurs directly from l to j, i.e., without visiting another state in between. Therefore, the transition hazards can be envisaged as momentary forces of transition between states l and j.

8.2 Multistate models as arising from a nested series of competing risks experiments

We now give an algorithmic description of a multistate model. The algorithm describes how the transition hazards (2.28) generate a multistate process via a nested series of competing risks experiments. This may be used to simulate multistate data, using competing risks simulation as in Section 3.2 as a key building block. We emphasize, however, that the algorithmic perspective goes beyond the question of implementing simulations. The algorithm addresses the more fundamental question of how a multistate process can be constructed. As with competing risks, this point of view is helpful to interpret analyses of multistate data. In this context, it is important to point out that the algorithm, building on a *nested series* of experiments, follows a time-dynamic perspective. Finally, the algorithm outlines why competing risks-type techniques should be available to analyse the transition *hazards* of a multistate model. Estimation of the transition *probabilities* will require some further work, however.

This is the algorithm. Consider an individual which is in state l at time 0, where l is some state in the state space $\{0, 1, 2, \ldots, J\}$. Recall that different individuals may start in different states at time 0. Given the individual's initial state l, the waiting time in state l and the state entered on leaving l are generated as follows.

1. The waiting time in state l is generated with hazard $\alpha_{l.}(t) = \sum_{j=0, j \neq l}^{J} \alpha_{lj}(t)$, $t \geq 0$.
2. Say that the waiting time generated in step 1 equals t_0. Then, the state X_{t_0} entered at this time is determined in a multinomial experiment, which decides with probability $\alpha_{lj}(t_0)/\alpha_{l.}(t_0)$ on state j, $j \neq l$.

Note that these two simulation steps are as in the competing risks case of Section 3.2. Also note that $\alpha_{lj}(t) = 0$ for pairs of states (l, j), for which an $l \to j$ transition is not modelled.

Say that the individual has entered state j at time t_0. If state j is absorbing, i.e., if there are no transitions out of state j, the algorithm stops. Otherwise, a second competing risks experiment follows.

1. The waiting time in state j is generated with hazard $\alpha_{j.}(t) = \sum_{\tilde{j}=0, \tilde{j} \neq j}^{J} \alpha_{j\tilde{j}}(t)$, $t \geq t_0$.
2. Say that the waiting time generated in step 1 equals t_1. Then, the state $X_{t_0+t_1}$ entered at this time is determined in a multinomial experiment, which decides with probability $\alpha_{j\tilde{j}}(t_0 + t_1)/\alpha_{j.}(t_0 + t_1)$ on state $\tilde{j}$, $\tilde{j} \neq j$.

Say that the individual has entered a transient (i.e., a non-absorbing) state $\tilde{j}$ at time $t_0 + t_1$. Then, another competing risks experiment is carried out, now starting at time $t_0 + t_1$ in state $\tilde{j}$. Further competing risks experiments are carried out until reaching an absorbing state. A process constructed along these lines will be a multistate model as described in Section 8.1. Readers interested in a thorough mathematical treatment are referred to Section 4.4 of Gill and Johansen (1990). An implementation of the algorithm is in the function mssample of the R package mstate. Exercise 3 in Section 4.6 discusses practical handling of hazards where the distribution of the corresponding waiting time does not spend 100% of the probability mass in $[0, \infty)$. The latter may, e.g., happen for empirical hazards as a consequence of censoring. We now focus on a number of important implications of the algorithm, which we exemplarily discuss for the illness-death model.

Consider the representation of the illness-death model as in Figure 8.1 (left) and, for the time being, assume that recovery is not modelled. We consider an individual, which is 'healthy' at time 0: the individual is in state 0 at time 0. For this individual, the algorithm starts with a standard competing risks experiment with transition hazards $\alpha_{01}(t)$ and $\alpha_{02}(t)$. Say that the individual leaves state 0 at time t_0. If the individual enters state 2 on leaving state 0 (i.e., $X_{t_0} = 2$), the algorithm stops.

If, however, $X_{t_0} = 1$, a second experiment follows. For the illness-death model of Figure 8.1 (left), this experiment is a degenerated competing risks experiment, because there is only one transition modelled out of state 1. A simple survival experiment is carried out, generating the individual's waiting time in state 1 with hazard $\alpha_{12}(t)$, $t \geq t_0$. Note, however, two important facts about this second experiment. First, the experiment starts at time t_0 of entry into state 1, i.e., we only consider $\alpha_{12}(t)$ for values t with $t \geq t_0$. Second, the time-inhomogeneous Markov property is relevant: the transition 'force' $\alpha_{12}(t)$ which acts upon the individual does depend on time t since time origin 0; the process is time-inhomogeneous. But it does *not* depend on the time t_0 of entry into state 1; the process is Markov.

The second step of the experiment immediately generalizes to competing endpoints, say, absorbing states 2 and $\tilde{2}$. Then a true competing risks experiment is being carried out, generating the individual's waiting time in state 1 with hazards $\alpha_{12}(t)$ and $\alpha_{1\tilde{2}}(t)$, $t \geq t_0$. This can be further extended to modelling 'recovery', i.e., allowing for $1 \to 0$ transitions. In an illness-death model *with* recovery and two competing absorbing states 2 and $\tilde{2}$, the individual is exposed to a competing risks experiment with hazards $\alpha_{10}(t)$, $\alpha_{12}(t)$ and $\alpha_{1\tilde{2}}(t)$, $t \geq t_0$. This implies that an individual may move back and forth between states 'healthy' and 'ill' a number of times before reaching an absorbing state. In other words, the individual may undergo several competing risks experiments in both states 0 and 1, but there is always only one competing risks experiment acting upon the individual at a single point in time.

Because the multistate model is generated by a nested sequence of competing risks experiments, the cumulative transition hazards should be estimable as in Section 4.1. Assuming, for the time being, a completely observed multistate process, e.g., a natural estimator for the cumulative hazard $A_{12}(t) = \int_0^t \alpha_{12}(u)\,du$ is

$$\widehat{A}_{12}(t) = \sum_{k=1}^{K} \frac{\text{number of observed } 1 \to 2 \text{ transitions at } s_k}{\text{number of individuals in state 1 just prior to } s_k}, \qquad (8.6)$$

where the sum is over all transition times s_k, $s_k \leq t$. The estimator is similar in structure to the cause-specific Nelson-Aalen estimators (4.8), but there are two important differences which reflect the more complex multistate structure.

First, individuals who are 'healthy', i.e., in state 0 of Figure 8.1 (left), at time 0 but later become 'ill' are only counted in the denominator of (8.6) right after their transition into state 1. We have already encountered risk set entries after time 0 for left-truncated data. In Section 4.4, pregnant women entered the study and hence the risk set after conception, which is time 0 for the duration of pregnancy. The present delayed entry into the risk set is, however, of a different nature. We assumed that the multistate process was completely observed. Therefore, individuals have been under observation since time 0. However, at time 0, 'healthy' individuals are in state 0 at time origin. They are at risk for making a transition out of state 0, but not for making a $1 \to 2$ transition. An originally 'healthy' individual only becomes at risk for making a $1 \to 2$ transition, *after* having moved from the 'healthy' state 0 into the 'illness' state 1.

Formally, this delayed entry into the risk set of being in state 1 is also a form of left-truncation, but it is important to remember that the delayed entry is not due to restrictions on the observable data, but rather generated by the multistate process itself. Therefore, it is sometimes called *internal* left-truncation (Andersen et al., 1993, p. 156–157), in contrast to external left-truncation, which is imposed by the observation scheme. On the other hand, realizing the similarity to (external) left-truncation implies that methods accounting for left-truncation as in Chapters 4 and 5 will be available for analysing the transition hazards.

The second difference arises in an illness-death model *with* recovery. In such a model, an individual who moves back and forth between the 'healthy' state and the 'illness' state will enter and drop out of the risk set (i.e., the denominator of (8.6)) several times before eventually making a transition into the absorbing state 2. Analogously, such an individual will even contribute to the numerator of the appropriate estimator of $A_{01}(t)$ as often as the individual makes a $0 \rightarrow 1$ transition.

The fact that an individual may contribute more than one observed transition of the same type sometimes raises dependency concerns in that the individual contributes a 'cluster of dependent data' to the analysis. The concern disappears if we follow the time-dynamic perspective of the generating algorithm. Here, the individual is never at risk for more than one competing risks experiment. It must be re-emphasized, however, that our treatment so far relies on the time-inhomogeneous Markov assumption. We return to this issue in Chapter 12.

8.3 Exercises

1. For a multistate model as in Section 8.1 show that

$$\mathbf{P}(s,t) = \mathbf{P}(s,u) \cdot \mathbf{P}(u,t)$$

   for $s \leq u \leq t$.

2. Simulate an illness-death model *with* recovery with 200 individuals that start in state 0 with a probability of 0.6. Transition hazards are defined as
   - $\alpha_{01}(t) = \frac{1}{3}t^2 + \frac{1}{5}$,
   - $\alpha_{02}(t) = \frac{3}{t+1}$,
   - $\alpha_{10}(t) = 1$,
   - $\alpha_{12}(t) = 2\sqrt{t}$.

   Compare the Nelson-Aalen estimator (8.6) against the true quantity.

9

Nonparametric estimation

9.1 The Nelson-Aalen estimator and the Aalen-Johansen estimator

We consider n individuals under study with individual multistate processes $(X_t^{(i)})_{t \geq 0}$, $X_t^{(i)} \in \{0, 1, 2, \ldots, J\}$, $i = 1, 2, \ldots n$. We assume that the n processes are, conditional on the initial states $X_0^{(i)}$, independent replicates of a multistate process as in Section 8.1. Observation of the individual multistate data is subject to a right-censoring time C_i and possibly also to a left-truncation time L_i. We assume that right-censoring and left-truncation are independent as explained in Section 2.2.2. The setting is similar to the competing risks situation of Section 4.1, but the individual data are now potentially more complex. We analogously introduce some notation connected to occupation of states in the model and possible transitions between them:

Let

$$Y_{l;i}(t) := \mathbf{1}(X_{t-}^{(i)} = l, L_i < t \leq C_i), \, l \in \{0, 1, 2, \ldots, J\}, \tag{9.1}$$

denote whether individual i is in state l and under observation just before time t. Here, we have written $X_{t-}^{(i)}$ for the value of the ith multistate process just before time t. We have $Y_{l;i}(t) = 1$ for individuals i who are in state l and under observation just before time t, and $Y_{l;i}(t) = 0$ otherwise. If $Y_{l;i}(t) = 1$, individual i may be *observed* to move out of state l at time t, or individual i may be censored at t or remain under observation in state l. We also write

$$N_{lj;i}(t) :=$$
individual i's number of observed direct $l \rightarrow j$ transitions in $[0, t]$, (9.2)

$l, j \in \{0, 1, 2, \ldots, J\}$, $l \neq j$. Here, 'direct $l \rightarrow j$ transition' means a transition from state l into state j without visiting another state in between. That is, the transition is made *directly* along the arrow that points from state l to state j. $N_{lj;i}(t)$ is a counting process.

We aggregate over all individuals $i = 1, \ldots, n$: the number of individuals to be observed at risk in state l just prior to time t is

$$Y_l(t) := \sum_{i=1}^{n} Y_{l;i}(t), \tag{9.3}$$

and the number of observed direct $l \to j$ transitions during the time interval $[0, t]$ is

$$N_{lj}(t) := \sum_{i=1}^{n} N_{lj;i}(t), \; l \neq j. \tag{9.4}$$

We also write

$$\Delta N_{lj}(t) := N_{lj}(t) - N_{lj}(t-) \tag{9.5}$$

for the increments of $N_{lj}(t)$, i.e., the number of $l \to j$ transitions observed *exactly* at time t.

We now motivate the Nelson-Aalen estimator of $A_{lj}(t) = \int_0^t \alpha_{lj}(u)\,\mathrm{d}u$ analogously to our derivation of the cause-specific Nelson-Aalen estimators in Section 4.1. Recall from Equation (8.4) that $\alpha_{lj}(t)\,\mathrm{d}t$ is an infinitesimal conditional transition probability

$$\alpha_{lj}(t) \cdot \mathrm{d}t = \mathrm{P}(X_{(t+\mathrm{dt})-} = j \mid X_{t-} = l).$$

If we observe no $l \to j$ transition at t (i.e., $\Delta N_{lj}(t) = 0$), we estimate the increment $\alpha_{lj}(t)\mathrm{d}t$ of the cumulative $l \to j$ hazard as 0. If we do observe $l \to j$ transitions at t (i.e., $\Delta N_{lj}(t) > 0$), we estimate this conditional transition probability as the ratio of the number $\Delta N_{lj}(t)$ of $l \to j$ transitions divided by the number $Y_l(t)$ at risk just prior to the transition time t. Summing up over these increments yields the Nelson-Aalen estimators:

$$\widehat{A}_{lj}(t) := \sum_{s \leq t} \frac{\Delta N_{lj}(s)}{Y_l(s)}, \; l \neq j, \tag{9.6}$$

where summation in (9.6) is over all observed event times in $[0, t]$. An estimator of the variance of $\widehat{A}_{lj}(t)$ is

$$\widehat{\sigma}_{lj}^2(t) := \sum_{s \leq t} \frac{\Delta N_{lj}(s)}{Y_l^2(s)}, \; l \neq j, \tag{9.7}$$

where summation is again over all observed event times in $[0, t]$. Approximate $100 \cdot (1 - \alpha)\%$ confidence intervals of $\widehat{A}_{lj}(t)$ at a given time point t, $\alpha \in (0, 1)$, can be constructed analogously to the competing risks case of Equation (4.10).

We now turn to probability estimation, which is seen to be more involved than estimating $A_{lj}(t)$. The derivation of the Nelson-Aalen estimators has been in analogy to their cause-specific counterparts. The analogy is built on the fact that multistate models are generated by a sequence of competing risks experiments, which are regulated by the transition hazards. We reiterate, however, that the present Nelson-Aalen estimators reflect the more complex multistate structure in that an individual may conceptually drop in and out of the state-specific risk sets several times. Similarly, an individual may be observed to have more than one direct $l \to j$ transition, although there is only at most one such transition at a fixed point in time. However, the transition probabilities are in general a complex function of the transition hazards, because the state occupied at some time t may potentially result from a complex nested series of competing risks experiments. In general, there may also be more than one possible sequence of competing risks experiments leading to being in a certain state at a certain time. For competing risks, we found that the probability estimates were deterministic functions of the Nelson-Aalen estimators. The same turns out to be true for the situation at hand, which should not come as a surprise given the generating algorithm. The key issue is to find the right mapping which connects probabilities and hazards.

In Section 2.2.4, we found that the matrix of transition probabilities $\mathbf{P}(s, t)$ can be approximated based on a partition $s = t_0 < t_1 < t_2 < \ldots < t_{K-1} < t_K = t$ of the time interval $[s, t]$ as

$$\mathbf{P}(s, t) \approx \prod_{k=1}^{K} \left(\mathbf{I} + \Delta\mathbf{A}(t_k) \right). \tag{9.8}$$

In (9.8), $\mathbf{I}$ is the $(J+1) \times (J+1)$ identity matrix, the (l, j)th element of $\Delta\mathbf{A}(t_k)$ is $A_{lj}(t_k) - A_{lj}(t_{k-1})$, and $A_{ll}(t) = -\sum_{j=0, j\neq l}^{J} A_{lj}(t)$. Computing the approximation for ever finer partitions of $[s, t]$ approaches a limit, namely a matrix-valued product integral $\prod$, which equals the matrix of transition probabilities,

$$\mathbf{P}(s, t) = \prod_{u \in (s,t]} \left(\mathbf{I} + \mathrm{d}\mathbf{A}(u) \right). \tag{9.9}$$

We do not repeat the derivation of approximation (9.8) here, but emphasize the consequences for estimation. The key issues are that product integration is the mapping that switches from cumulative transition hazards to the matrix of transition probabilities, that *all* cumulative transition hazards are involved, and that plugging in the Nelson-Aalen estimators in (9.9) results in a finite matrix product as in (9.8). The latter is the important Aalen-Johansen estimator (Aalen and Johansen, 1978)

$$\widehat{\mathbf{P}}(s, t) = \prod_{u \in (s,t]} \left(\mathbf{I} + \mathrm{d}\widehat{\mathbf{A}}(u) \right), \tag{9.10}$$

which is an ordinary, finite matrix product over all event times u in $(s, t]$ and matrices $\mathbf{I} + \mathrm{d}\widehat{\mathbf{A}}(u)$. Here, we have written $\widehat{\mathbf{A}}(u)$ for the matrix of Nelson-Aalen estimators with (l, j)th entry $\widehat{A}_{lj}(u)$ as in Equation (9.6) for $l \neq j$, $\widehat{A}_{ll}(u) :=$

$-\sum_{j,j\neq l}\widehat{A}_{lj}(u)$, and $\mathrm{d}\widehat{\mathbf{A}}(u)$ for the matrix with entries $\widehat{A}_{lj}(u)-\widehat{A}_{lj}(u-)$. The Aalen-Johansen estimator is also called the empirical transition matrix, which explains the name of the R package etm.

The estimator (9.10) is written in very compact form. We investigate it in some more detail. To begin, assume that t_1 is the first observed event time after s. Then

$$\widehat{\mathbf{P}}(s,t_1) = \mathbf{I} + \mathrm{d}\widehat{\mathbf{A}}(t_1).$$

The lth row of the estimator contains the estimates $\mathrm{P}_{lj}(s,t_1)$. For $j \neq l$, this probability is estimated as the number of observed $l \to j$ transitions at time t_1 divided by the number of individuals observed to be in state l at time s and, hence, just prior to time t_1. The diagonal element is such that the sum over the lth row equals 1. This latter fact justifies the definition $\widehat{A}_{ll}(u) := -\sum_{j,j\neq l}\widehat{A}_{lj}(u)$ below Equation (9.10).

Next, assume that t_2 is the first observed event time after t_1. Then,

$$\widehat{\mathbf{P}}(s,t_2) = \widehat{\mathbf{P}}(s,t_1) \cdot \widehat{\mathbf{P}}(t_1,t_2)$$
$$= \left(\mathbf{I} + \mathrm{d}\widehat{\mathbf{A}}(t_1)\right)\left(\mathbf{I} + \mathrm{d}\widehat{\mathbf{A}}(t_2)\right).$$

Consider the (l,j)th entry, where l may be equal to j,

$$\widehat{\mathrm{P}}\left(X_{t_2} = j \,|\, X_s = l\right) = \sum_{\tilde{j}=0}^{J}\widehat{\mathrm{P}}(X_{t_1} = \tilde{j} \,|\, X_s = l) \cdot \widehat{\mathrm{P}}(X_{t_2} = j \,|\, X_{t_1} = \tilde{j})$$

$$= \sum_{\tilde{j}=0}^{J}\left(\mathbf{1}(\tilde{j} = l) + \Delta\widehat{A}_{l\tilde{j}}(t_1)\right) \cdot \left(\mathbf{1}(\tilde{j} = j) + \Delta\widehat{A}_{\tilde{j}j}(t_2)\right).$$

It is worthwhile to point out that the first line in the previous display again relies on the Markov property. Because conditioning on the state occupied at t_1 is tantamount to conditioning on the entire previous history, including the state occupied at s, the estimated quantity is

$$\sum_{\tilde{j}=0}^{J}\mathrm{P}(X_{t_1} = \tilde{j} \,|\, X_s = l) \cdot \mathrm{P}(X_{t_2} = j \,|\, X_{t_1} = \tilde{j}, X_s = l)$$

$$= \sum_{\tilde{j}=0}^{J}\mathrm{P}(X_{t_2} = j, X_{t_1} = \tilde{j} \,|\, X_s = l) = \mathrm{P}(X_{t_2} = j \,|\, X_s = l).$$

Finally, the terms $(\mathbf{1}(\tilde{j} = l) + \Delta\widehat{A}_{l\tilde{j}}(t_1)) \cdot (\mathbf{1}(\tilde{j} = j) + \Delta\widehat{A}_{\tilde{j}j}(t_2))$ are products of simple proportions as explained earlier.

Although the Aalen-Johansen estimator (9.10) is algebraically simple when computed in a step-by-step, entry-by-entry fashion, it must be noted that, in general, no closed formulae are available. However, for some comparatively simple, but practically quite important models closed formulae do exist. If available, closed expressions may be derived by solving a Kolmogorov forward differential equation (e.g., Andersen et al., 1993, Section II.6). We do not go into detail here, but prefer to exemplarily focus on the *interpretation* of such a formula for the illness-death model without recovery. There are parallels between the present case and our interpretation of the cumulative incidence functions (3.11) and the corresponding Aalen-Johansen estimator (4.18), and there are differences which highlight the more complex structure of the illness-death model.

To begin, consider the probabilities $P_{00}(s,t) = P(X_t = 0 \,|\, X_s = 0)$ and $P_{11}(s,t) = P(X_t = 1 \,|\, X_s = 1)$ to stay in state 0 or 1 until time t, given that the individual had already been in the respective state at time s, $s \le t$. These probabilities are

$$P_{00}(s,t) = \prod_{u \in (s,t]} \big(1 - (\alpha_{01}(u) + \alpha_{02}(u))d(u)\big)$$
$$= \exp\left(-\int_s^t \alpha_{01}(u) + \alpha_{02}(u)d(u)\right)$$

and

$$P_{11}(s,t) = \prod_{u \in (s,t]} \big(1 - \alpha_{12}(u)d(u)\big)$$
$$= \exp\left(-\int_s^t \alpha_{12}(u)d(u)\right).$$

These quantities are essentially common survival probabilities, either with all-cause hazard $\alpha_{01} + \alpha_{02}$ or with one single hazard α_{12}, taking time s as time origin. Their Aalen-Johansen estimators $\widehat{P}_{00}(s,t)$ and $\widehat{P}_{11}(s,t)$ are equal to the corresponding Kaplan-Meier estimators. Note that this survival-type situation is naturally implied by the algorithmic competing risks perspective of Section 8.2.

Next, consider the probability $P_{01}(s,t) = P(X_t = 1 \,|\, X_s = 0)$,

$$P_{01}(s,t) = \int_s^t P_{00}(s,u-)\alpha_{01}(u)P_{11}(u,t)\,du.$$

The first two terms in the integrand of the preceding display are as for the cumulative incidence function (3.11). There, our interpretation was that one integrates or, loosely speaking, sums up over 'infinitesimal probabilities' of making the $0 \to 1$ transition exactly at time u. The difference now is that we also have to include $P_{11}(u,t)$ to ensure that an individual also stays in state 1 until time t after having made the $0 \to 1$ transition at time u. The reason for this is that the illness-death model also models transitions out of state 1, whereas state 1 was absorbing in the competing risks case. The Aalen-Johansen estimator is

$$\widehat{P}_{01}(s,t) = \sum_{u,s<u\leq t} \widehat{P}_{00}(s,u-) \cdot \frac{\Delta N_{01}(u)}{Y_0(u)} \cdot \widehat{P}_{11}(u,t).$$

Using $P_{02}(s,t) = 1 - P_{00}(s,t) - P_{01}(s,t)$, we also find that

$$P_{02}(s,t) = \int_s^t P_{00}(s,u-)\alpha_{02}(u)\,du$$

$$+ \int_s^t P_{00}(s,u-)\alpha_{01}(u)P_{12}(u,t)\,du,$$

where $P_{12}(u,t) = 1 - P_{11}(u,t)$. Interpretation and estimation of the preceding display are analogous to $P_{01}(s,t)$.

Before turning to examples, we must finally comment on (co-)variance estimation for the Aalen-Johansen estimator. As with standard survival data and competing risks, a Greenwood-type variance estimator will generally be preferred. Such an estimator does exist, but, unfortunately, the algebra is becoming increasingly complex. In their Section IV.4, Andersen et al. (1993) develop a recursion formula for such an estimator, and the estimator is implemented in R. Further R-specific details are reported in de Wreede et al. (2010) and Allignol et al. (2011a). The covariance estimator times n, the number of individuals under study, is an asymptotically consistent estimator of the covariance matrix of the limit process of

$$t \mapsto \sqrt{n}\left(\widehat{\mathbf{P}}(s,t) - \mathbf{P}(s,t)\right), s \leq t.$$

As the limit process is Gaussian with mean zero, this can again be used to construct pointwise confidence intervals in the usual way. Rather than going into technical details here, we illustrate variance estimation and construction of confidence intervals in the examples to follow.

9.2 Examples

9.2.1 Impact of hospital-acquired pneumonia on length of stay and mortality in intensive care units

The data set icu.pneu is included in the kmi package and is described in detail in Chapter 1. The data set contains information on 1313 patients. 21 observations were censored. 108 patients experienced hospital-acquired pneumonia. Of these, 82 patients were discharged alive and 21 patients died. Without prior hospital-acquired pneumonia, 1063 patients were discharged alive and 126 patients died.

The data are an example of an illness-death model without recovery and with competing endpoints. There is one common initial state entered by all patients on admission. Occurrence of hospital-acquired pneumonia is modelled by transitions into the intermediate state. The competing endpoints are death and alive discharge. Recovery is not modelled (and also difficult to diagnose in practice). This means that being in the infectious state reflects past exposure, but not necessarily present infection status. The subject matter interest is to compare patients who have acquired pneumonia in hospital with those who have not.

The aim of our analysis is twofold: first, we investigate the impact of hospital-acquired pneumonia (HAP) on length of intensive care unit stay. In hospital epidemiology, length of stay is often used to quantify healthcare costs. Increased length of stay is used in cost benefit studies to justify further infection control measures. This analysis does not distinguish between the competing endpoint states, but must account for the time dependency of HAP. Not accounting for this time dependency will typically lead to artificially inflated results; see also Section 11.3 on the so-called 'time-dependent bias'. Second, we investigate the impact of HAP on intensive care unit mortality. This analysis must account for both the time dependency of infection status and for competing endpoints.

Combined endpoint analysis.

We consider an illness-death model as in Figure 8.1 (left). Here, state 0 is entered on admission to the unit. State 1 is reached when a patient acquires an infection and state 2 is the combined end-of-stay endpoint. This is illustrated in Figure 9.1.

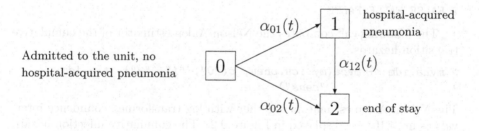

Fig. 9.1. Illness-death model without recovery: investigating the impact of hospital-acquired pneumonia (HAP) on length of intensive care unit stay.

We start by computing the cumulative transition hazards using the mvna package. The first step is to transform the data into a format that is suitable for using the mvna function. This step is similar to the competing risks case in Section 4.2 and left to readers as an exercise. An excerpt of the resulting data set, which we call my.icu.pneu, is displayed below.

```
     id entry exit from to to2        age sex
7    405     0    1    0  2   3 84.69435   F
8    410     0    6    0  1   1 69.80130   M
9    410     6   28    1  2   3 69.80130   M
21  3163     0    5    0  2   2 48.19945   M
89 17743     0   19    0  1   1 71.84670   F
90 17743    19   21    1  2   2 71.84670   F
93 17776     0    7    0  2   2 90.99458   F
```

Entry times into and exit times out of a state are contained in `entry` and `exit`, respectively. The state from which a transition occurs is in `from`. The state reached is in `to` and `to2`. The difference between the two columns is that `to` does not distinguish between the competing endpoints. We use `to` in this first analysis. In both `to` and `to2`, censored observations are indicated by 'cens'. We now define the matrix indicating the possible transitions. The possible transitions are from state 0 to states 1 or 2, and from state 1 to state 2.

```
> tra.idm <- matrix(FALSE, 3, 3,
+                   dimnames = list(c(0, 1, 2), c(0, 1, 2)))
> tra.idm[1, 2:3] <- TRUE
> tra.idm[2, 3] <- TRUE
> tra.idm

      0     1     2
0 FALSE  TRUE  TRUE
1 FALSE FALSE  TRUE
2 FALSE FALSE FALSE
```

The following call computes the Nelson-Aalen estimator of the cumulative transition hazards.

```
> mvna.idm <- mvna(my.icu.pneu, c("0", "1", "2"), tra.idm,
+                   "cens")
```

The Nelson-Aalen estimators together with log-transformed confidence intervals as in (4.10) are displayed in Figure 9.2. The cumulative infection hazard is seen to be low in comparison to the other hazards. This means that most patients do not acquire pneumonia during intensive care unit stay. However, those who do acquire pneumonia have an increased length of stay, because the cumulative hazard for end of stay out of the infectious state is less than the one out of the initial state.

We now estimate the transition probabilities using the `etm` package.

```
> etm.idm <- etm(my.icu.pneu, c("0", "1", "2"), tra.idm,
+                "cens", s = 0)
```

We begin by plotting the transition probability $P_{01}(0, t)$. In cancer research, this probability is sometimes called the 'probability of being in response function' (Temkin, 1978), if the intermediate state of the illness-death model is

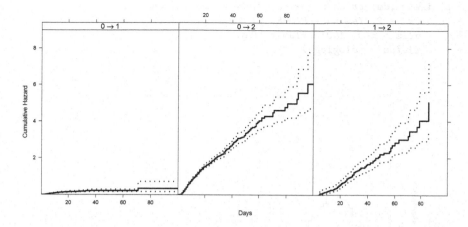

Fig. 9.2. *Hospital-acquired pneumonia data.* Nelson-Aalen estimators: The cumulative infection hazard is in the left plot. The middle plot is for end of stay without prior infection. The cumulative hazard for end of stay after prior infection is in the right plot.

interpreted as 'response to treatment'. In the present example, $P_{01}(0,t)$ is the probability to still be in the unit after having acquired infection.

A display of the Aalen-Johansen estimator of $P_{01}(0,t)$ along with pointwise confidence intervals is in Figure 9.3. Specification of the transition probability of interest is done through `tr.choice == '0 1'` and confidence intervals are displayed setting the argument `conf.int` to `TRUE`. By default, a confidence interval without transformation is displayed by the `plot` function. Coverage is often improved by using a transformation. A complementary log-log transformation as used in (4.21) can be specified with the option `ci.fun = 'cloglog'`. The Aalen-Johansen estimator along with confidence intervals can easily be recovered through the `summary` function, for instance, using `summary(etm.idm)$'0 1'[, c('P', 'lower', 'upper')]`. Transformation of the confidence intervals are handled via the `ci.fun` argument. Figure 9.3 shows that the proportion of HAP patients increases for about the first 20 days and then decreases again.

In practice, specification of `ci.fun` may be complicated by the fact that different types of transformations may be used depending on the transition probability at hand. For instance, a log-log transformation as in (4.16) may be used for the state occupation probabilities, but a complementary log-log transformation may be applied for actual transitions. `ci.fun` accounts for this by accepting vector arguments. Transition probabilities are returned by the `summary` function following a specific order. First, the actual transitions are returned in lexicographical order. Next come the conditional state occupation probabilities in lexicographical order. In the HAP example:

```
> plot(etm.idm, tr.choice = "0 1", conf.int = TRUE,
+      lwd = 2, legend = FALSE, ylim = c(0, 0.1),
+      xlim = c(0, 100), xlab = "Days",
+      ci.fun = "cloglog")
```

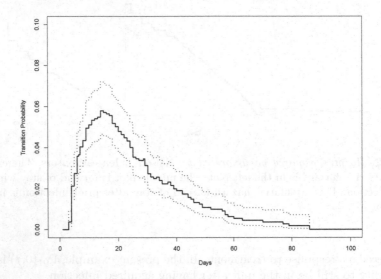

Fig. 9.3. *Hospital-acquired pneumonia data.* Aalen-Johansen estimator of $P_{01}(0, t)$ with pointwise 95% confidence intervals based on a complementary log-log transformation.

```
> names(summary(etm.idm))
```

```
[1] "0 1" "0 2" "1 2" "0 0" "1 1"
```

The transformations of the confidence intervals are made following the same order. In order to have a complementary log-log transformation for the actual transitions and a log-log transformation on the conditional state occupation probabilities, we would use `ci.fun = c(rep('cloglog', 3), rep('log', 2))`.

The construction of the confidence intervals is based on a Greenwood-type estimator of the covariance matrix of the Aalen-Johansen estimator; see Andersen et al. (1993, Section IV.4.1.3) for a detailed account. The Greenwood-type estimator has been found to be the preferred estimator for both single endpoint survival data and competing risks, and we therefore also recommend it for use with more complex multistate models. The estimator is implemented in `etm`. In the present example, the estimator is contained in `etm.idm$cov`. A detailed explanation of how to extract estimated variances and covariances from `etm.idm$cov` has been given in Section 4.2; see, in particular, Table 4.2.

The aim of the present analysis has been to ascertain the effect of HAP on the length of intensive care unit stay. The Nelson-Aalen estimators of Figure 9.2 showed that HAP patients have a prolonged stay. Figure 9.3 illustrated the proportion of HAP patients over the course of time. In addition to these plots, *probability* plots that illustrate such a prolonging effect are often requested in practice. The challenge is that HAP status is not a baseline condition. As a consequence, simple Kaplan-Meier-type plots comparing, in the present example, infected versus non-infected are not available; see also Section 11.3.

Therefore, Anderson et al. (1983) suggested the 'landmark method'. The idea is to select a range of landmark time points s. Given HAP status at time s, we then compare the probabilities of having reached the absorbing state by time t, $s \leq t$. A typical application of the landmark approach would be to take subsets of the data conditional on HAP status at time s. Next, probability estimates would be computed within HAP groups defined at time s, taking s as the new time origin. In the present multistate setting, we can also consider the Aalen-Johansen estimators of $P_{02}(s,t)$ and $P_{12}(s,t)$ for different times s. For that, we need to compute the transition probability matrix for the different starting times and, for convenience, put the resulting etm objects in a list.

```
> time.points <- c(seq(3, 10, 1), 15)
> landmark.etm <- lapply(time.points, function(start.time) {
+     etm(my.icu.pneu, c("0", "1", "2"), tra.idm, "cens",
+         start.time)
+ })
```

The Aalen-Johansen estimators of $P_{02}(s,t)$ and $P_{12}(s,t)$ are displayed in Figure 9.4 for a range of landmarks s. The prolonging effect of HAP is illustrated by the fact that, overall, $\widehat{P}_{02}(s,t) \geq \widehat{P}_{12}(s,t)$. The effect is most pronounced for early landmarks.

Analysis of competing endpoints in a progressive model.

We now turn to the competing endpoint analysis, using a progressive illness-death model as in Figure 8.1 (right) but with competing absorbing states. This model has two transient states and four absorbing states. As before, the initial state 0 is entered on admission, occurrence of HAP is modelled by transitions into the intermediate state 1. The difference between the previous model and the present model is that there are states 2 and 3 for death and discharge, respectively, which can only be reached from state 1. And there are states 4 and 5 for death and discharge, respectively, which can only be reached from state 0. In other words, being either in state 2 or in state 4 represents having died in the unit. Patients in state 2 acquired HAP before death, whereas patients in state 4 did not. This is illustrated in Figure 9.5.

The aim of the present analysis is to study the effect of HAP on intensive care unit mortality. We need a slightly transformed data set and a new matrix

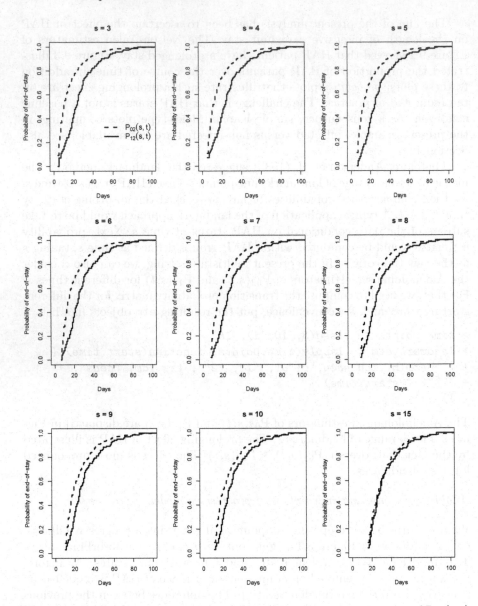

Fig. 9.4. *Hospital-acquired pneumonia data.* Aalen-Johansen estimators of $P_{02}(s,t)$ (dashed lines) and $P_{12}(s,t)$ (solid lines) for different landmark times s.

of transition probabilities, reflecting the multistate model at hand. An excerpt of the new data set, which we call `my.icu.pneu.prog`, is presented below.

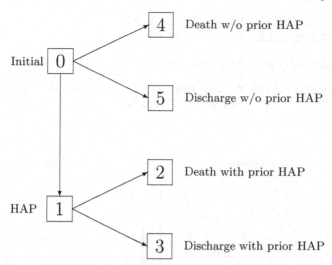

Fig. 9.5. Progressive illness-death model without recovery and competing endpoints for the occurrence of hospital-acquired pneumonia (HAP).

	id	entry	exit	from	to	to2	age	sex
7	405	0	1	0	5	3	84.69435	F
8	410	0	6	0	1	1	69.80130	M
9	410	6	28	1	3	3	69.80130	M
21	3163	0	5	0	4	2	48.19945	M
89	17743	0	19	0	1	1	71.84670	F
90	17743	19	21	1	2	2	71.84670	F
93	17776	0	7	0	4	2	90.99458	F

The only change is in to, which is 2 or 3 for death and discharge, respectively, when an individual has been previously infected, and 4 or 5 for death and discharge, otherwise. We are using the recoded to in the present analysis.

We define the matrix of possible transitions

```
> tra.prog <- matrix(FALSE, 6, 6,
+        dimnames = list(as.character(0:5), as.character(0:5)))
> tra.prog[1, c(2, 5:6)] <- TRUE
> tra.prog[2, 3:4] <- TRUE
> tra.prog

        0     1     2     3     4     5
0 FALSE  TRUE FALSE FALSE  TRUE  TRUE
1 FALSE FALSE  TRUE  TRUE FALSE FALSE
2 FALSE FALSE FALSE FALSE FALSE FALSE
3 FALSE FALSE FALSE FALSE FALSE FALSE
```

```
4 FALSE FALSE FALSE FALSE FALSE FALSE
5 FALSE FALSE FALSE FALSE FALSE FALSE
```

and estimate the cumulative transition hazards.

```
> mvna.prog <- mvna(my.icu.pneu.prog, as.character(0:5),
+                   tra.prog, "cens")
```

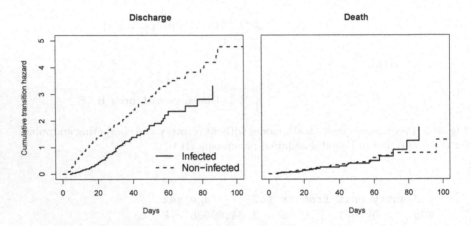

Fig. 9.6. *Hospital-acquired pneumonia data.* Nelson-Aalen estimator of the cumulative transition hazards of discharge (left) and death (right), either starting from the infectious state (solid lines) or the initial state (dashed lines).

Figure 9.6 displays the Nelson-Aalen estimators of the cumulative transition hazards into the absorbing states. Interestingly, the interpretation is similar to the analysis of pneumonia *on admission* in Section 4.3: HAP increases mortality via a reduced hazard for alive discharge, whereas the hazard for death in the unit remains essentially unchanged.

The following call computes the matrix of transition probabilities.

```
> etm.prog <- etm(my.icu.pneu.prog, as.character(0:5),
+                 tra.prog, "cens", 0)
```

Figure 9.7 displays the estimated *state occupation probabilities* $\widehat{P}(X_t = j)$, $j = 2, \ldots, 5$. Because all individuals are initially in state 0, we have that $P(X_t = j) = P_{0j}(0, t)$, and the state occupation probabilities are conveniently estimated by the Aalen-Johansen estimator.

Figure 9.7 is *not* well suited to illustrate the effect of HAP on intensive care unit mortality. The state occupation probabilities for states 2 and 3 are low in comparison, because the HAP hazard had been found to be low in Figure 9.2. An important information obtained from the figure is that the

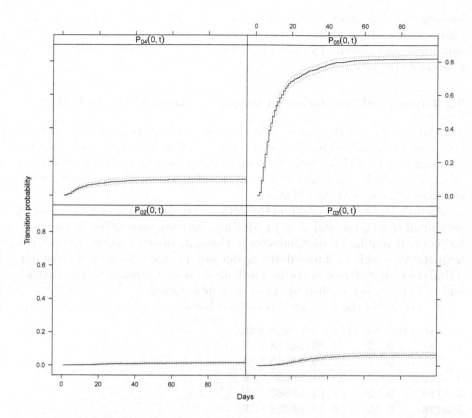

Fig. 9.7. *Hospital-acquired pneumonia data.* Aalen-Johansen estimators of $P_{0j}(0, t)$, $j = 2, \ldots, 5$. The upper panel is for patients without prior HAP, the lower panel is for previously infected patients. The right panel is for alive discharge; the left panel is for death in intensive care unit.

vast majority of patients is discharged alive without prior infection (top right plot). Comparing the right panel with the left panel of the Figure, we also find that patients are more likely to be discharged than to die, regardless of the infection status.

Probability plots that illustrate the increase of mortality after HAP are again a challenging issue due to the time dependency of the infection status. One approach would be the landmarking method as used in the previous analysis, but now accounting for the competing endpoints. That is, we might consider plotting $\hat{P}_{0j}(s, t)$ and $\hat{P}_{1j}(s, t)$ for a range of landmarks s. We also note that Schumacher et al. (2007) considered attributable mortality after a time-dependent exposure such as HAP and related probability plots.

We finally reiterate that the present calculations relied on a Markov assumption. In Section 11.3, we discuss how the Markov assumption can be investigated. We also note that Allignol et al. (2011b) assessed the Markov assumption for the hospital-acquired pneumonia data from which the present subsample was drawn. The authors argued that the Markov property may reasonably be assumed.

9.2.2 Impact of ventilation on length of intensive care unit stay

The data set `sir.cont` is included in the `mvna` package as described in Chapter 1 and contains information on times of ventilation and time of intensive care unit stay for 747 patients. The aim of the present analysis is to study the effect of ventilation on length of stay in the unit. We do so by estimating the cumulative hazards for end of stay.

An important characteristic of the data is that ventilation may be switched on and off during hospital stay. In addition, patients may either be on ventilation or off ventilation on admission to the unit. In other words, the present multistate model is an illness-death model *with* recovery. There is no common initial state. State 0 represents 'no ventilation', state 1 represents 'ventilation', and end of stay is modelled by transitions into state 2.

An excerpt of the data set is presented below.

```
  id from to time       age sex
1   41    0  2    4 75.34153   F
2  395    0  2   24 19.17380   M
3  710    1  0   33 61.56568   M
4  710    0  2   37 61.56568   M
5 3138    0  2    8 57.88038   F
6 3154    0  2    3 39.00639   M
```

`sir.cont` is already formatted for using the `mvna` or `etm` packages: `from` is the state from which a transition occurs, `to` is the state to which a transition occurs, and `time` is the transition time.

We define the matrix specifying the possible transitions below.

```
> tra.ventil <- matrix(FALSE, 3, 3, dimnames =
+                    list(c("0", "1", "2"), c("0", "1", "2")))
> tra.ventil[1, c(2, 3)] <- TRUE
> tra.ventil[2, c(1, 3)] <- TRUE
> tra.ventil

      0     1     2
0 FALSE  TRUE  TRUE
1  TRUE FALSE  TRUE
2 FALSE FALSE FALSE
```

and estimate the cumulative transition hazards using the `mvna` function.

```
> mvna.ventil <- mvna(sir.cont, c("0", "1", "2"),
+                    tra.ventil, "cens")
```

Figure 9.8 displays the Nelson-Aalen estimators $\widehat{A}_{02}(t)$ and $\widehat{A}_{12}(t)$ with log-transformed confidence intervals as in (4.10). Ventilation is seen to prolong length of stay.

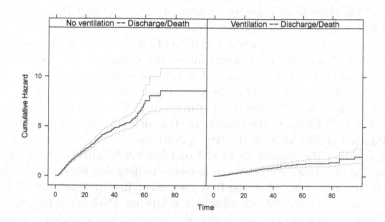

Fig. 9.8. *Ventilation data.* Nelson-Aalen estimates for the transitions no ventilation $\rightarrow$ end of stay (left) and ventilation $\rightarrow$ end of stay (right).

9.3 Exercises

1. Simulate an illness-death model *with* recovery with 200 individuals that start in state 0 with a probability of 0.6. Transition hazards are defined as
   - $\alpha_{01}(t) = \frac{1}{3}t^2 + \frac{1}{5}$,
   - $\alpha_{02}(t) = \frac{3}{t+1}$,
   - $\alpha_{10}(t) = 1$,
   - $\alpha_{12}(t) = 2\sqrt{t}$.

   Simulate random censoring times following a uniform distribution leading to about 20% of the observations being censored.

   Estimate the cumulative transition hazards for the simulated data using the mvna package. Compare the estimated quantities with the true quantities.

2. Cox models for the transition hazards can typically be fitted in 'univariate fashion', using standard Cox software; see Chapter 10. To give a foretaste of the principles needed to fit Cox models in the multistate framework, estimate $A_{12}(t)$ in 'univariate fashion':

a) Use the mvna function, considering the $1 \to 2$ transition as the only possible transition. That is, entry times will be entry times into state 1, and transitions into state 0 are considered as censored observations.

b) Use the output of the survfit function of the survival package, following the strategy of Exercise 2a.

3. Using the data from Exercise 1, estimate $P_{02}(s,t)$ and $P_{12}(s,t)$ for $s = 0.2, 0.4, \ldots, 1.2$, and do the landmark plots of Figure 9.4.

4. Resimulate the illness-death model data of Exercise 1 with *state-dependent censoring*. Instead of considering latent censoring times, consider a 'censoring state'. From state 0, an individual could move to state 1 or 2 with the hazards defined earlier or to a censoring state with hazard 0.9. From state 1, the hazard to observe a transition to the 'censoring state' is defined as $t^{1/3}$. Estimate the cumulative transition hazards and compare the estimated quantities with the true quantities.

5. We consider the multistate model of Figure 9.9. This model is used to describe the treatment course of patients undergoing bone marrow transplant for leukemia. After the first treatment line, patients are at risk of dying due to treatment side effects or relapsing. Patients in relapse might be offered a salvage therapy (state 4), namely donor lymphocyte infusion (DLI), that proved out to be successful in restoring remission of the disease (state 6). A more thorough explanation of the model can be found in Klein et al. (2000) and in Allignol et al. (2011a). The aim of the analysis is to quantify the effectiveness of both bone marrow transplant and DLI. The current leukemia-free survival probability (CLFS) (Klein et al., 2000) may be used as a summary measure of the model. CLFS corresponds to patients alive and leukemia-free, either in the first or second post-transplant remission. That is, CLFS is the probability to be in state 0 or 6 at time t. The nature of the present multistate model as a nested series of competing risks experiments is clearly seen in the figure. We assume that the model is Markov.

a) Simulate 500 individuals stemming from the multistate process of Figure 9.9 with constant hazards defined as:
   - $\alpha_{01} = 0.4$,
   - $\alpha_{02} = 0.5$,
   - $\alpha_{23} = 1.3$,
   - $\alpha_{24} = 0.9$,
   - $\alpha_{45} = 0.5$,
   - $\alpha_{46} = 1.4$,
   - $\alpha_{67} = 0.05$,
   - $\alpha_{68} = 0.07$.

b) Estimate and plot the CLFS which is a functional of the matrix of the transition probabilities and is estimated as

$$\widehat{\mathrm{CLFS}}(t) = \widehat{P}_{00}(0,t) + \widehat{P}_{06}(0,t),$$

and variance estimator equal to

$$\widehat{\text{VAR}}(\widehat{\text{CLFS}}(t)) =$$
$$\widehat{\text{VAR}}(\widehat{P}_{00}(0,t)) + \widehat{\text{VAR}}(\widehat{P}_{06}(0,t)) + 2\widehat{\text{cov}}(\widehat{P}_{00}(0,t), \widehat{P}_{06}(0,t)).$$

c) Redo the analysis with a random sample of the real data that can be found at `http://www.jstatsoft.org/v38/i04/supp/3`.

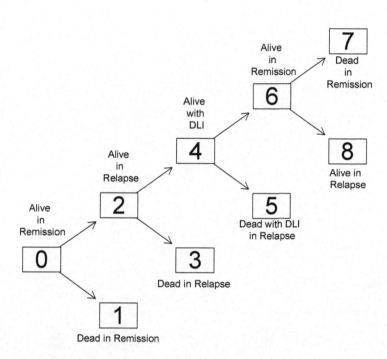

Fig. 9.9. Multistate model used for Exercise 6. DLI stands for donor lymphocyte infusion.

10

Proportional transition hazards models

10.1 Model formulation and practical implementation

As with competing risks, the most widely used regression model for multistate data assumes a proportional hazards form for the transition hazards of the multistate model. We re-emphasize that the proportional hazards assumption is made for interpretational and technical convenience. As in Chapter 9, we consider n individuals under study with individual multistate data subject to independent right-censoring and/or left-truncation. This entails that right-censoring and left-truncation may depend on covariates included in the model. The n multistate processes are assumed to be conditionally independent given the baseline covariate values and given the initial states.

Consider the $l \to j$ transition hazard as in (8.4), but now related to individual covariate information,

$$\alpha_{lj;i}(t; Z_i) = \alpha_{lj;0}(t) \cdot \exp\left(\beta_{lj} \cdot Z_i\right), \; l, j \in \{0, 1, 2, \ldots, J\}, \, l \neq j, \, i = 1, \ldots, n, \tag{10.1}$$

where β_{lj} is a $1 \times p$ vector of regression coefficients, Z_i is a $p \times 1$ vector of covariates for individual i, and $\alpha_{lj;0}(t)$ is an unspecified, non-negative baseline hazard function. In principle, different transition hazards may be related to different covariate vectors. We also write

$$A_{lj;0}(t) = \int_0^t \alpha_{lj;0}(u)\mathrm{d}u \quad \text{and} \quad A_{lj;i}(t; Z_i) = \int_0^t \alpha_{lj;i}(u; Z_i)\mathrm{d}u \tag{10.2}$$

for the respective cumulative transition hazards.

The model formulation is in obvious analogy to that for the cause-specific hazards of a competing risks model in (5.1). There, in Section 5.1, we reformulated the models for the cause-specific hazards in terms of one vector β, which contained all regression coefficients and did not depend on the transition type, and transition-specific covariate vectors. Such a reformulation allows some covariates to have a common effect on some hazards, and it may be analogously achieved for the present multistate case. For ease and brevity of presentation, we, however, mainly focus on formulation (10.1). We briefly return to the model reformulation in Chapter 12, where we comment on model parsimony. The necessary data set preparation is also briefly illustrated in the practical examples of Section 10.2.

Besides the analogy of (10.1) to proportional cause-specific hazards models, there are also important differences, which are related to our discussion of the Nelson-Aalen estimator following Equation (8.6). To be specific, assume that $l = 1$ and $j = 2$ in an illness-death model. That is, we model the transition hazard of 'death' after 'illness'. Regardless of whether we model recovery (i.e., $1 \to 0$ transitions) the model

$$\alpha_{12;i}(t; Z_i) = \alpha_{12;0}(t) \cdot \exp\left(\beta_{12} \cdot Z_i\right), \ i = 1, \ldots, n, \qquad (10.3)$$

reflects the Markov assumption (8.3). In the preceding display, the hazard for making a 'diseased' $\to$ 'dead' transition does depend on time t since time origin and on baseline, i.e., time-fixed covariates. However, the hazard does *not* depend on the time at which a 'healthy' individual has entered the 'illness' state. There was no such restriction in the case of competing risks, because all individuals are in the single transient state of a competing risks model at time 0. It is a remarkable fact that hazard models such as in (10.1) can also incorporate dependency on *time-dependent* covariates such as entry time into state 1. Among other things, this would allow us to relax the Markov assumption by, e.g., modelling dependence of the $1 \to 2$ hazard on the sojourn time in the 'illness' state. However, time-dependent covariates pose some interpretational challenges, which can be best addressed from a multistate perspective, and we postpone a discussion to Chapter 11.

The second important difference becomes apparent in an illness-death model *with* recovery, where $1 \to 0$ transitions are feasible. Model (10.3) assumes that the $1 \to 2$ hazard does *not* depend on how often one has been ill or has recovered before. Model (10.1) expressed for the 'getting ill' hazard,

$$\alpha_{01;i}(t; Z_i) = \alpha_{01;0}(t) \cdot \exp\left(\beta_{01} \cdot Z_i\right),$$

assumes a common baseline hazard $\alpha_{01;0}(t)$ for successive $0 \to 1$ transitions. In other words, it is assumed that the baseline hazard for 'getting ill' is the same both for individuals who have not been 'ill' before, and for individuals who acquired 'illness' in the past but have recovered. Again, this assumption may be relaxed by including time-dependent covariates that carry information on the number of times one has been 'ill' before or on how long one has been 'ill' in the past. Another approach would be to extend the underlying illness-death model to different illness and/or health states such as 'ill for the first time', 'ill for the second time', 'first recovery', 'second recovery' and so on.

A key fact now is that models such as (10.1) may be fitted using standard Cox software *if* we take care of the observations made above. The same principle has been demonstrated for the algebraically simpler Nelson-Aalen estimator in Exercise 2 of Section 9.3. The idea is understood best via an example. Consider the following hypothetical individual 1 who moves through an illness-death model with recovery:

State occupied at t	For times t in...	Health status
$X_t^{(1)} = 0$	$t \in [0,3)$	'healthy'
$X_t^{(1)} = 1$	$t \in [3,5)$	'ill'
$X_t^{(1)} = 0$	$t \in [5,6)$	'healthy'
$X_t^{(1)} = 1$	$t \in [6,10)$	'ill'
$X_t^{(1)} = 2$	$t \geq 10$	'dead'

The individual falls 'ill' at time 3, 'recovers' at time 5, is 'ill' again from times 6 through 10, and finally 'dies' at time 10. Recall that individuals are not per se required to be 'healthy' at time origin. E.g., an individual who is 'ill' from the start, but 'recovers' at time 5 would have the first two lines in the preceding display replaced by $X_t^{(1)} = 1$ for $t \in [0,5)$.

The data set corresponding to individual 1 *for the analysis of the* $0 \rightarrow 1$ *hazard* would be:

```
id from to  entry exit  status
1   0    1   0     3     1
1   0    1   5     6     1
```

In the preceding display, status equal to 1 indicates that a $0 \rightarrow 1$ transition has been observed. The data set for the analysis of the $0 \rightarrow 2$ hazard has different values for status, but is otherwise identical:

```
id from to  entry exit  status
1   0    2   0     3     0
1   0    2   5     6     0
```

The interpretation is that individual 1 has been *at risk* for making a $0 \rightarrow 2$ transition during the time intervals $(0,3)$ and $(5,6)$, but such a transition has not been observed.

This is the data set contribution of individual 1 for analysing the $1 \rightarrow 0$ hazard:

```
id from to  entry exit  status
1   1    0   3     5     1
1   1    0   6     10    0
```

Note that a $1 \rightarrow 0$ transition has been observed at time 5, and that individual 1 was again at risk for making such a transition during the time interval $(6,10)$, but that the individual did *not* move into state 0 at time 10.

Finally, the data set contribution for analysing the $1 \rightarrow 2$ hazard is:

```
id from to  entry exit  status
1   1    2   3     5     0
1   1    2   6     10    1
```

If, say, the aim is an analysis of the $1 \rightarrow 2$ hazard, we must create a data set which consists of data lines as in the preceding display for each individual.

Only individuals who have entered state 1 at some point in time are part of the data set. Additional columns of the data set will contain the covariates. Then, the theory as outlined in Section 5.2.1 goes through, treating entry times into state 1 as 'internal' left-truncation and transitions out of state 1 into some state j, $1 \neq j \neq 2$ as, formally, right-censoring. (In the present example, only $j = 0$ must be considered.) We consider worked examples in Section 10.2.

Among the first to notice that Cox-type models (10.1) can be fitted using standard procedures were Kalbfleisch and Prentice (1980) (see their Section 7.3) and Kay (1982). In fact, already Cox, in his seminal paper (Cox, 1972), also considered bivariate survival times. Kalbfleisch and Prentice also outlined the analogy to methods for competing risks; see page 183 of Kalbfleisch and Prentice (1980). Their presentation starts from a sequence of event times and event types, whereas our process formulation is similar to Kay (1982). It should be noted, though, that these two representations of the data are equivalent (e.g., Gill and Johansen, 1990, Section 4.4), the relation being the data-generating algorithm of Section 8.2, and that the 2002 edition of the book by Kalbfleisch and Prentice also adopts a multistate process point of view (Kalbfleisch and Prentice, 2002, Section 8.3). A definite theoretical treatment has been given by Andersen and Borgan (1985), which also covers the case of common covariate effects; see also Andersen et al. (1993, Section VII.2).

Despite these early references, it was arguably the book by Therneau and Grambsch (2000) which, backed by actual computer code (mostly S-Plus), popularized Cox-type analyses of transitions hazards; see, in particular, their Section 8.6. Interestingly, most of their analyses use a robust variance estimator (see also our Section 5.4), which is motivated by the fact that a single individual may contribute more than one event to the analysis. So far, we have argued that *if* model (10.1) is correct, the standard procedures work, including variance estimation. However, we have also argued that model misspecification may be more of a concern here as compared to competing risks data, if, e.g., we assume a common baseline hazard for all currently 'healthy' individuals, regardless of their prior 'health' status. This has also been pointed out by Kay (1982). We investigate this matter in Sections 10.2 and 10.3.

Finally, we explain how transition probabilities may be predicted based on models (10.1). This is the general idea: the cumulative baseline hazards $A_{lj;0}(t)$ may be estimated in analogy to the methods for competing risks. For competing risks, we had to analyse all cause-specific hazards in order to subsequently predict probabilities. In the multistate context, we must analyse the transition hazards for all transitions $l \rightarrow j$, $l \neq j$. In the last step, product integration as in Section 9.1 is used to move from hazards to probabilities; see in particular Equation (9.10).

To be specific, recall from (9.1) that $Y_{l;i}(t)$ denotes whether individual i is at risk of an observed transition out of state l at time t and define the weighted risk set

$$S_{lj}^{(0)}(\beta_{lj}, t) := \sum_{i=1}^{n} \exp\left(\beta_{lj} \cdot Z_i\right) \cdot Y_{l;i}(t), \; l, j \in \{0, 1, 2, \ldots, J\}, \, l \neq j, \quad (10.4)$$

cf. Equation (5.8) for the case of competing risks. A Breslow estimator of the cumulative $l \to j$ baseline hazard analogous to (5.18) then is

$$\widehat{A}_{lj;0}(t) := \sum_{s \leq t} \frac{\Delta N_{lj}(s)}{S_{lj}^{(0)}(\widehat{\beta}_{lj}, s)}, \; l, j \in \{0, 1, 2, \ldots, J\}, \, l \neq j. \quad (10.5)$$

In (10.5), $\widehat{\beta}_{lj}$ is the vector of estimated regression coefficients, which is obtained from fitting a Cox model for the $l \to j$ transition as explained earlier. Summation in (10.5) is over all event times s, $s \leq t$. The predicted cumulative $l \to j$ hazard under model (10.1) is

$$\widehat{A}_{lj}(t; z) = \widehat{A}_{lj;0}(t) \cdot \exp\left(\widehat{\beta}_{lj} \cdot z\right), \, l \neq j. \quad (10.6)$$

For the aim of predicting probabilities, we also define

$$\widehat{A}_{ll}(t; z) := - \sum_{j, j \neq l} \widehat{A}_{lj}(t; z) \quad (10.7)$$

(cf. the nonparametric analog below Equation (9.10)). Also recall that different transitions $l \to j$ and $l \to \tilde{j}$, $j \neq \tilde{j}$ may be related to different covariates, although we have not made this explicit in the preceding displays. In (10.7), this may be formally achieved by making z large enough, such that all necessary covariate information is included. Components of z, which are not part of the original model for a single transition, will then have their regression coefficients set to zero for that transition. We then predict the matrix of transition probabilities by replacing the nonparametric $\widehat{A}_{lj}(t)$ in the Aalen-Johansen formula (9.10) with $\widehat{A}_{lj}(t; z)$,

$$\widehat{P}(s, t; z) = \prod_{u \in (s, t]} \left(I + d\widehat{A}(u; z)\right). \quad (10.8)$$

In (10.8), $\widehat{A}(u; z)$ is the matrix with (l, j)-entry $\widehat{A}_{lj}(t; z)$, and the covariate vector z is again 'large enough'.

Similar to prediction for competing risks data and to (co-)variance estimation for the Aalen-Johansen estimator, estimating variances and covariances is becoming algebraically increasingly complex, such that we refrain from going into technical details here. Readers are referred to Section VII.2.3 of Andersen et al. (1993) for a thorough theoretical development and to de Wreede et al. (2010) for further R-specific details. We again focus on practical implementation in the examples to follow. We also note that, as a computational alternative, the bootstrap may also be used to obtain variances and covariances; see Section 2.3 and Appendix A.

10.2 Examples

10.2.1 Hospital-acquired pneumonia

We consider the icu.pneu data that we have analysed in Section 9.2.1 with a combined endpoint reflecting end of intensive care unit stay. The aim of the present analysis is to study the effect of age and sex on each of the transition hazards using Cox proportional hazards models. For this, we use the recoded data set my.icu.pneu as in Section 9.2.1. Fitting a proportional hazards model using the coxph function can be done in several ways. The first one is to use a subset of the data set containing the people at risk of making the transition of interest, as explained in Section 10.1. The second possibility is to fit one Cox model to an extended data set as in Section 5.2.2, stratifying on the transition type.

We start with the first method. The following calls compute Cox models for the $0 \to 1$ and $0 \to 2$ transitions, respectively.

```
> cox.icu.pneu01 <- coxph(Surv(entry, exit, to == 1) ~ age +
+                         sex, my.icu.pneu, subset = from == 0)
> cox.icu.pneu02 <- coxph(Surv(entry, exit, to == 2) ~ age +
+                         sex, my.icu.pneu, subset = from == 0)
```

Subsetting is achieved using subset = from == 0. Results are displayed below.

```
> summary(cox.icu.pneu01)

  n= 1313

         coef exp(coef) se(coef)     z Pr(>|z|)
age  0.007949  1.007981 0.005631 1.412    0.158
sexM 0.210054  1.233744 0.201131 1.044    0.296

     exp(coef) exp(-coef) lower .95 upper .95
age      1.008     0.9921    0.9969     1.019
sexM     1.234     0.8105    0.8318     1.830

> summary(cox.icu.pneu02)

  n= 1313

         coef exp(coef)  se(coef)      z Pr(>|z|)
age  0.001658  1.001659  0.001632  1.016   0.3098
sexM -0.110789  0.895128  0.058937 -1.880   0.0601

     exp(coef) exp(-coef) lower .95 upper .95
age     1.0017     0.9983    0.9985     1.005
sexM    0.8951     1.1172    0.7975     1.005
```

Older age has a non-significant increasing effect on both transition hazards out of state 0. Males have a higher hazard of infection, but a lower hazard of directly leaving the unit; these effects are not significant, either.

We now fit the model for the $1 \to 2$ transition. This time, we select individuals at risk in state 1 using subset = from == 1.

```
> cox.icu.pneu12 <- coxph(Surv(entry, exit, to == 2) ~ age +
+                         sex, my.icu.pneu, subset = from == 1)
> summary(cox.icu.pneu12)

  n= 108


          coef exp(coef)  se(coef)       z Pr(>|z|)
age  -0.010061  0.989990  0.007147 -1.408    0.159
sexM  0.012586  1.012665  0.205342  0.061    0.951


     exp(coef) exp(-coef) lower .95 upper .95
age      0.990     1.0101    0.9762     1.004
sexM     1.013     0.9875    0.6771     1.514
```

We may also fit the proportional hazards models using an extended data set, if we stratify the analysis on the transition type. The idea has been illustrated for competing risks in Section 5.2.2. The trick is to extend the data set so that for each transient state an individual reaches, we have one line per possible transition from this transient state. In addition, there has to be a status variable specifying whether a specific transition has been observed.

As illustrated for competing risks, the extended data frame allows us to model a common effect of a covariate on some or all transition hazards. The extended data frame is also used for making predictions, using mstate. We exemplarily discuss the extended data frame, which we call my.icu.pneu.ext, for one individual. The extended data set has a new column new.to, which contains all states that can be reached from the state in from, and a new column new.status, which is 1 if the transition has been observed.

```
> my.icu.pneu.ext[my.icu.pneu.ext$id == 2010304, ]

            id entry exit from        age sex new.to new.status trans
1224  2010304     0   14    0 73.62053   M      1          1     1
12241 2010304     0   14    0 73.62053   M      2          0     2
12251 2010304    14   22    1 73.62053   M      2          1     3
            age.1     age.2    age.3 male.1 male.2 male.3
1224  73.62053   0.00000  0.00000      1      0      0
12241  0.00000 73.62053  0.00000      0      1      0
12251  0.00000  0.00000 73.62053      0      0      1
```

The extended data set is displayed above for individual 2010304. The first two lines are concerned with the transitions starting from state 0. This individual acquires pneumonia, hence new.status equals 1 when new.to is 1 (and is 0 when new.to is 2). The last line is for the $1 \rightarrow 2$ transition, which is the only possible transition out of state 1. We have also added a trans variable that indicates the transition type: 1 for transition $0 \rightarrow 1$, 2 for transition $0 \rightarrow 2$ and so on. age.h and male.h, $h = 1, 2, 3$, are transition-specific covariates that take their original value for transition h, and are set to 0 otherwise. We now fit the Cox model using the extended data set. We use new.status as the status indicator and stratify on the transition type, specifying strata(trans) in the formula.

```
> fit.cox.icu.ext <- coxph(Surv(entry, exit, new.status) ~
+                          age.1 + age.2 + age.3 +
+                          male.1 + male.2 + male.3 +
+                          strata(trans), my.icu.pneu.ext)
> summary(fit.cox.icu.ext)

  n= 2734

            coef exp(coef)  se(coef)      z Pr(>|z|)
age.1   0.007949  1.007981  0.005631  1.412   0.1580
age.2   0.001658  1.001659  0.001632  1.016   0.3098
age.3  -0.010061  0.989990  0.007147 -1.408   0.1592
male.1  0.210054  1.233744  0.201131  1.044   0.2963
male.2 -0.110789  0.895128  0.058937 -1.880   0.0601
male.3  0.012586  1.012665  0.205342  0.061   0.9511

        exp(coef) exp(-coef) lower .95 upper .95
age.1      1.0080     0.9921    0.9969     1.019
age.2      1.0017     0.9983    0.9985     1.005
age.3      0.9900     1.0101    0.9762     1.004
male.1     1.2337     0.8105    0.8318     1.830
male.2     0.8951     1.1172    0.7975     1.005
male.3     1.0127     0.9875    0.6771     1.514
```

Readers can check by looking at the estimated regression coefficients and at the confidence intervals for the estimated hazard ratios that the results are identical to those obtained from fitting separate models.

One usage of the extended data set is making predictions with the mstate package (de Wreede et al., 2010, 2011). As an illustration, we predict $P_{01}(0, t, z)$ for a woman and a man 60.14 years old, which is the median age in the sample. As in the competing risks example in Section 5.2.2, the first step is to define the matrix of possible transitions. mstate supplies the convenience function trans.illdeath for illness-death models.

```
> mat <- trans.illdeath()
> mat
```

```
          to
from      healthy illness death
  healthy     NA       1     2
  illness     NA      NA     3
  death       NA      NA    NA
```

We then create a data frame with covariate information for the two hypothetical individuals for whom we wish to make predictions. For convenience, we artificially create an **msdata** object, which allows us to use the **expand.covs** function for generating transition-specific covariates.

```
> woman.medianage <- data.frame(age = rep(median.age, 3), male
+                                 = rep(0, 3), trans = 1:3)
> attr(woman.medianage, "trans") <- mat
> class(woman.medianage) <- c("msdata", "data.frame")
> woman.medianage <- expand.covs(woman.medianage,
+                                   c("age", "male"))
> woman.medianage$strata <- 1:3
> man.medianage <- data.frame(age = rep(median.age, 3), male
+                               = rep(1, 3), trans = 1:3)
> attr(man.medianage, "trans") <- mat
> class(man.medianage) <- c("msdata", "data.frame")
> man.medianage <- expand.covs(man.medianage, c("age", "male"))
> man.medianage$strata <- 1:3
```

As explained in the competing risks example, it is, at the time of writing, safer to make the predictions starting from a Cox proportional hazards model that uses the Breslow method for handling ties. We thus refit the model and use the **msfit** function to obtain the predicted cumulative transition hazards for both individuals.

```
> fit.cox.icu.ext.bres <- coxph(Surv(entry, exit, new.status)
+                                 ~ age.1 + age.2 + age.3 +
+                                 male.1 + male.2 + male.3 +
+                                 strata(trans), my.icu.pneu.ext,
+                                 method = "breslow")
> msfit.woman <- msfit(fit.cox.icu.ext.bres, woman.medianage,
+                         trans = mat)
> msfit.man <- msfit(fit.cox.icu.ext.bres, man.medianage,
+                         trans = mat)
```

The transition probabilities are then predicted using the **probtrans** function.

```
> pt.woman <- probtrans(msfit.woman, 0)
> pt.man <- probtrans(msfit.man, 0)
```

For illustration, we exemplarily consider the probability to still be in the unit after having acquired infection; see also Section 9.2.1. The predicted probability $\widehat{P}_{01}(0, t; z)$ can be found in , e.g., pt.woman[[1]]$pstate2 and associated standard errors in pt.woman[[1]]$se2. The numbering of the components pstate and se is as explained in Section 5.2.2. Figure 10.1 displays $\hat{P}_{01}(0, t; z)$ for a woman and a man of median age. Overall, this probability is seen to be higher for men. Note the usefulness of the presentation in Figure 10.1: based

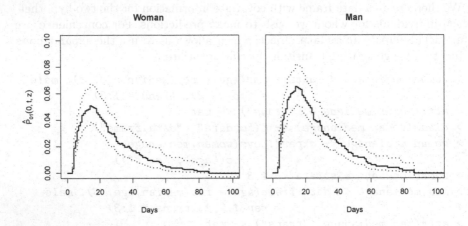

Fig. 10.1. *Hospital-acquired pneumonia data.* Predicted probability to still be in intensive care unit after having acquired infection. The left plot is for a woman of median age; the right plot is for a man of the same age. The dashed lines are 95% pointwise confidence intervals based on complementary log-minus-log transform.

on the Cox analyses of the transition hazards, a higher probability for a man might not be surprising, because male sex reduced the hazard for a $0 \to 2$ transition and increased the hazard for a $0 \to 1$ transition. This increases the probability that a man reaches the infectious state as compared to a woman. However, male sex also increased the hazard for a transition out of the infectious state. This decreases the probability that a man stays in the infectious state as compared to a woman. Figure 10.1 provides for a useful summary of these transition hazard-based findings, also accounting for the baseline transition hazards.

10.2.2 Ventilation in intensive care unit

We reconsider the data of Section 9.2.2. In Section 9.2.2, we found that being ventilated prolongs intensive care unit stay. We now exemplarily study the effect of age and sex on the hazards for the transition $0 \to 1$, i.e., switching ventilation on, and for the transition $1 \to 0$, i.e., switching ventilation off. We also compute robust variance estimates; see also Section 5.4. In the present setting, an individual may contribute more than one data line because ventilation can be switched on and off. Robust variance estimation is then performed

by `coxph`, if we specify a `cluster(id)` term in the formula. Here, `id` is the
variable which identifies the lines belonging to one person.

```
> cox.ventil.01 <- coxph(Surv(time, to == 1) ~ age + sex +
+                         cluster(id), sir.cont,
+                         subset = from == 0)
> cox.ventil.10 <- coxph(Surv(time, to == 0) ~ age + sex +
+                         cluster(id), sir.cont,
+                         subset = from == 1)
> summary(cox.ventil.01)

  n= 686

          coef exp(coef)  se(coef) robust se      z Pr(>|z|)
age   0.008977  1.009017  0.006683  0.005860  1.532    0.126
sexM -0.154848  0.856545  0.237356  0.214902 -0.721    0.471

      exp(coef) exp(-coef) lower .95 upper .95
age      1.0090      0.991    0.9975     1.021
sexM     0.8565      1.167    0.5621     1.305

> summary(cox.ventil.10)

  n= 455

          coef exp(coef)  se(coef) robust se      z Pr(>|z|)
age  -0.009433  0.990612  0.003321  0.003353 -2.813  0.00491
sexM -0.115315  0.891085  0.114944  0.115391 -0.999  0.31763

      exp(coef) exp(-coef) lower .95 upper .95
age      0.9906      1.009    0.9841    0.9971
sexM     0.8911      1.122    0.7107    1.1172
```

Robust estimates of the standard error are in the **robust se** column, and
the confidence intervals are now based on **robust se**. In the present example,
both the model-based estimator of the standard error and the robust estimator
are in close agreement. We note that the difference may be more pronounced
in other data examples. For comparison, we also report estimated standard
errors based on a nonparametric bootstrap. R code for the present bootstrap
analysis can be found in Appendix A.

```
      age01       sex01       age10       sex10
0.005932345  0.2247314  0.003404829  0.1168784
```

The standard errors have been computed based on 10000 bootstrap samples.
The values are close to those previously found. In general, we would expect
the nonparametric bootstrap results to be closer to the robust estimation
procedure, if there is a relevant difference between **se** and **robust se**.

10.3 Exercises

1. If model (10.1) is correct, standard procedures for fitting proportional hazards models apply, even if individuals contribute more than one event to the analysis. The major aim of the present exercise is to investigate variance estimation both under a correctly specified model and under a misspecified model in an illness-death model *with recovery*. We first simulate data following model (10.1). We then introduce model misspecification by increasing the baseline hazards after a first recovery.

a) Consider the multistate model of Figure 10.2. This is a modified illness-death model with recovery in which the state 'healthy after first recovery' is distinguished from the initial 'healthy' state. We define two scenarios. The first one reduces to the usual illness-death model with recovery: we set $\alpha_{01}(t) = \alpha_{21}(t)$ and $\alpha_{03}(t) = \alpha_{23}(t)$. The second scenario considers the full model. Baseline transition hazards for the first scenario are:

- $\alpha_{01}(t) = \alpha_{21}(t) = 1.2$,
- $\alpha_{03}(t) = \alpha_{23}(t) = 0.5$,
- $\alpha_{12}(t) = 0.7$,
- $\alpha_{13}(t) = 0.8$.

The second scenario has different hazards out of state 2:

- $\alpha_{21}(t) = 2$,
- $\alpha_{23}(t) = 0.6$.

Let Z be a binary baseline covariate. Data for $Z = 1$ stem from a perfect Cox model with regression coefficients (indexed by transition type) $\beta_{01} = \beta_{21} = 0.1$, $\beta_{03}(t) = \beta_{23} = 0.3$ and no effect on the other transitions. Generate 300 observations for $Z = 0$ and 300 observations for $Z = 1$ for both scenarios.

b) For both sets of data, analyse the effect of Z on the transitions from the healthy state assuming a usual illness-death model with recovery, i.e., without distinguishing between states 0 and 2. Compute both the usual and the robust variance estimates.

c) Repeat the analyses of the previous exercise, but this time distinguishing between states 0 and 2.

d) In a simulation study, repeat all previous steps 1000 times. Draw boxplots of the estimated regression coefficients. Compare the variance estimates with an estimator of the variance based on the distribution of the 1000 estimated regression coefficients.

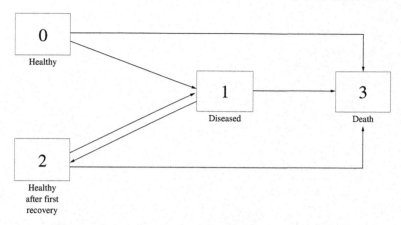

Fig. 10.2. Modified illness-death model for Exercise 1

11

Time-dependent covariates and multistate models

Our treatment of regression models for cause-specific or transition hazards in Chapters 5 and 10 has so far been restricted to baseline covariates Z only, those covariates whose value is measured or known at time origin. It is a remarkable strength of regression models for *hazards* that they can also incorporate covariates $Z(t)$ whose value may change with time. Although this extension is a good deal technically straightforward, new interpretational challenges arise. Some of these may conveniently be addressed from a multistate perspective, which is what we do in this chapter.

There are different types of time-dependent covariates. The future values of *defined* time-dependent covariates are known at time origin. E.g., if time 0 is treatment initiation, a patient's age at time t is a defined time-dependent covariate. Another example is covariate-time interactions such as $Z \cdot f(t)$, where f is a known function of time.

An important distinction, due to Kalbfleisch and Prentice (1980) who also give formal definitions, is between external and internal time-dependent covariates.

External covariates are, informally speaking, those covariates whose existence does not depend conceptually on the individual under study. A canonical example of an external covariate is air pollution in a study on asthma events (Kalbfleisch and Prentice, 1980, p. 123). The current level of air pollution may have an effect on the hazard of an asthma attack, but it is assumed that air pollution is not influenced by the individual's asthma event.

Internal covariates do conceptually depend on the individual under study. A common example is an individual's current health status, say $Z(t)$, either healthy or ill. The information that a patient is currently healthy (or ill) does imply that the patient is currently alive. In contrast, it is unclear what we *mean* by $Z(t)$, if the individual has died before t.

In this chapter, we do not aim at a comprehensive but inevitably incomplete account on time-dependent covariates. Instead, we focus on the connection between multistate models and time-dependent covariates, which are useful in many applications. Such covariates are typically internal. Our approach follows ideas such as found in Andersen (1986), Beyersmann and Schumacher (2008) and Cortese and Andersen (2010).

Section 11.1 discusses multistate models as simple joint model for time-dependent covariates and time-to-event endpoints. Proportional hazards models with time-dependent covariates are also introduced in this section. The reasoning applies quite generally to transition hazards as in Chapter 10. We specifically discuss competing risks in Section 11.2. Besides including time-dependent covariates in an analysis of the cause-specific hazards, we also discuss how to analyse such information using subdistribution hazards. This latter aspect particularly profits from approaching both time-dependent covariates and subdistribution processes from a multistate perspective. As an example, we investigate the impact of hospital-acquired pneumonia on length of stay and hospital outcome in intensive care unit.

Finally, Section 11.3 addresses some issues that are often raised in the analysis of time-dependent covariates.

11.1 A simple joint model for time-dependent covariates and time-to-event endpoints

The basic idea is to interpret a multistate model as a joint model for both the time-dependent covariate process and the time-to-event of interest. The time-dependent covariate reflects the transitions between the transient states of the multistate model, whereas time-to-event is modelled by the time until the multistate process enters an absorbing state. Among other things, this model is 'simple', because it conceptually covers time-dependent covariates with a finite range only. On the other hand, such a 'simple' model is applicable in many practical situations.

The connection between multistate models and time-dependent covariates is best understood via an example. We exemplarily discuss the illness-death model without recovery in Section 11.1.1, but the ideas immediately transfer to more complex models. An example of a proportional hazards model with time-dependent covariates is then discussed based on the idea of an underlying illness-death model in Section 11.1.2.

11.1.1 Transient states and time-dependent covariates

We exemplarily consider the situation of hospital-acquired pneumonia from Sections 9.2.1 and 10.2.1. For the time being, we focus on a combined endpoint 'hospital discharge', either alive or dead. Hence, we have a multistate process $(X_t)_{t \geq 0}$ with state space $\{0, 1, 2\}$. Every patient enters the initial state 0 on admission. Occurrence of hospital-acquired pneumonia (HAP) is modelled by transitions into the intermediate state 1 and hospital discharge is modelled by transitions into the absorbing state 2. The model is an illness-death model as in Figure 8.1 (left).

Let T denote the length of hospital stay,

$$T := \inf\{t > 0 \,|\, X_t = 2\}$$

with discharge hazard

$$\alpha(t)dt := P(T \in dt \,|\, T \geq t).$$

The aim is to relate the discharge hazard to a time-dependent covariate $Z(t)$ which reflects HAP status. A standard requirement is that $t \mapsto Z(t)$ is left-continuous such that the value of $Z(t)$ is known just before time t. The interpretation of this requirement is that the covariate information *just before t* is related to the hazard of an event *at t*.

By writing $Z(t)$ or $t \mapsto Z(t)$, we tacitly assume that $Z(t)$ denotes a covariate which is conceptually well defined for any choice of t. Recall, however, from the introduction of this Chapter that this may not be the case. It is not evident what we *mean* by 'HAP status' or by $Z(t)$ for times t *after* end of hospital stay, $t > T$. For definiteness, we identify $Z(t)$ with

$$t \mapsto Z(t \wedge T),$$

the covariate process *stopped* at the event time T. The interpretation of the preceding display is that only the information up to the event time will be related to the hazard of the event. For internal covariates, i.e., covariates whose existence conceptually depends on the individual under study, looking at the stopped covariate process also ensures that we need not speculate about hypothetical covariate values such as 'hospital-acquired pneumonia after end of hospital stay'. It should be noted, though, that only the covariate values before stopping are used to relate the hazard to the covariate. This is so, because $\alpha(t)dt = P(T \in dt \,|\, T \geq t)$, and the condition in $P(T \in dt \,|\, T \geq t)$ requires that the event has not happened yet.

In our example, we define $Z(t)$ to be

$$Z(t) = \begin{cases} 1 & : \quad \text{if the patient has acquired pneumonia in } (0, t \wedge T), \\ 0 & : \quad \text{otherwise.} \end{cases}$$

The interpretation is that $Z(t) = 1$, if HAP has been acquired before t. Note that $Z(t)$ is left-continuous. Also note that $Z(t)$ does not reflect current infection status, but past exposure.

This is the connection of $Z(t)$ to the illness-death model X_t: Basically, the time-dependent covariate reflects the transitions between the transient states of the multistate model. Figure 11.1 illustrates the mechanism for an individual who acquires pneumonia on day 4 of hospital stay and whose hospital stay ends on day 10. The multistate process initially equals 0, $X_t = 0$ for all $t \in [0, 4)$. At time 4, the multistate process moves into the intermediate state 1 (i.e., $X_4 = 1$). The process stays in the intermediate state until end

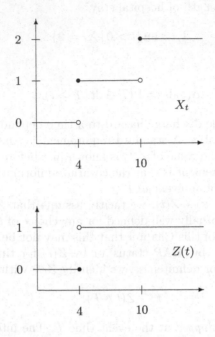

Fig. 11.1. Top: Multistate process X_t for an individual who acquires HAP on day 4 and who is discharged on day 10. Bottom: Time-dependent covariate $Z(t)$ reflecting HAP status. Bullets • are included in the graph, but circles ∘ are not.

of hospital stay at time 10, $X_t = 1$ for all $t \in [4, 10)$. At time 10, the multistate process moves into the absorbing state 2, i.e., $X_{10} = 2$, and stays in state 2 thereafter. This is a typical data set entry for analysis with mvna or etm, assuming the individual to have id 1:

```
id from  to  time
 1   0    1    4
 1   1    2   10
```

The multistate process and the time-dependent covariate are equal on the time intervals $[0, 4)$ and $(4, 10)$. Because $Z(t)$ is left-continuous, the covariate does not change its value precisely at time 4, but only 'a moment' later. This is illustrated in Figure 11.1. The difference is minimal: the individual can only make an 'HAP → end-of-stay' transition after time 4 both in terms of the multistate process and from the point of view of the covariate. $Z(t)$ does not change its value at time 10 of hospital discharge or thereafter. This reflects that no transitions between transient states occur for times $[10, \infty)$. The corresponding data set entry for a *Cox analysis* is in the following Section 11.1.2.

11.1.2 A proportional hazards model with time-dependent covariates

A proportional hazards model, which relates the individual HAP status of Section 11.1.1 to the discharge hazard, is now given by

$$\alpha_i(t; Z_i(t)) = \alpha_0(t) \cdot \exp\left(\beta \cdot Z_i(t)\right), \, i = 1, \ldots, n, \qquad (11.1)$$

where $Z_i(t)$ is individual i's time-dependent HAP status as in Figure 11.1 (bottom) and $\alpha_0(t)$ is the baseline discharge hazard. In principle, further co-variates could be included, but we stick to the simple model (11.1) for ease of presentation. We, however, illustrate how to include further covariates in the Example of Section 11.1.3.

There are two basic but important remarks to be made on model (11.1). First, the model essentially assumes the transition hazards into the absorbing state of the illness-death model in Figure 8.1 (left) to be proportional. The ratio of the transition hazards is estimated to be $\exp \widehat{\beta}$, i.e., the estimated hazard ratio which is obtained from fitting (11.1). Second, model (11.1) as such does not allow us to predict probabilities, because it does not provide for an analysis of the $0 \rightarrow 1$ hazard in the illness-death model.

Discussing these issues in a little more detail, note that model (11.1) states that $\alpha_0(t)$ is the discharge hazard for individuals without prior exposure to HAP. This is the transition hazard $\alpha_{02}(t)$ in the illness-death model. The discharge hazard for individuals with prior exposure to HAP then is $\alpha_{12}(t)$. Hence, model (11.1) states that

$$\frac{\alpha_{12}(t)}{\alpha_{02}(t)} = \exp(\beta).$$

We illustrate below how to fit the model. Practically, the key step is to generate an appropriate data set, which may then be analysed by, e.g., coxph. The data set contribution by individual 1 from the previous Section 11.1.1 would be:

```
id start  stop  z.t  status
1    0     4     0     0
1    4    10     1     1
```

These data set lines are somewhat different from the ones in the previous section that we would have used to estimate the cumulative transition hazards, say, and therefore deserve a comment. To begin, we have entries start and stop, which obviously play the role of entry and exit in a multistate context. We could safely use the names entry and exit instead of start and stop in the data lines above, but we have chosen not to do so. The coding in the previous display is sometimes referred to as start stop-notation. It invokes the idea of the beginning of a time interval and the end of a time interval. The difference between the time intervals $[0, 4)$ and $[4, 10)$ is that the covariate z.t reflecting HAP status $Z(t)$ changes its value from 0 to 1. In contrast, the names `entry` and `exit` invoke the idea of moving in and out of a state in a multistate model. We reiterate that the difference between these names does not truly matter. Finally, `status` informs on the fact that the event of interest (i.e., hospital discharge) has not been observed by the end of the first time interval, but that hospital discharge has happened at the end of the second time interval.

Using an appropriate data set in conjunction with `coxph` provides for an estimator $\widehat{\beta}$. In addition, an estimator of the baseline hazard $\alpha_{02}(t)$ may be obtained in the usual way, and this also allows for predicting $\alpha_{12}(t)$ under model (11.1). However, model (11.1) does *not* make a statement about the infection hazard $\alpha_{01}(t)$. We have, however, seen in, e.g., Chapter 9 that the transition probabilities of a multistate model in general depend on *all* transition hazards, including $\alpha_{01}(t)$. *As a consequence, predicting probabilities based on model (11.1) alone is in general meaningless.* This is quite intuitive. Assume that infection prolongs hospital stay, $\alpha_{12}(t) < \alpha_{02}(t)$ or, equivalently under model (11.1), $\exp\beta < 1$. Imagine two different situations, one where the infection hazard is small compared to $\alpha_{02}(t)$, the other where the infection hazard is large in comparison. Only the infection hazard differs between these two scenarios. Length of hospital stay tends to be smaller in the first scenario.

11.1.3 Example

Impact of hospital-acquired pneumonia (HAP) on length of stay in intensive care unit

We reconsider the data set `icu.pneu` from Sections 9.2.1 and 10.2.1. The data set is already formatted for fitting a Cox model with the HAP status as a time-dependent covariate. Time-dependent HAP status is in `pneu`. The following call fits a Cox model with HAP status as a time-dependent covariate.

```
> cox.hap.tdp <- coxph(Surv(start, stop, status) ~ pneu,
+                        icu.pneu)
> summary(cox.hap.tdp)

  n= 1421


          coef exp(coef) se(coef)      z Pr(>|z|)
pneu1  -0.4223    0.6556   0.1064 -3.969 7.23e-05


       exp(coef) exp(-coef) lower .95 upper .95
pneu1     0.6556      1.525    0.5322    0.8076
```

HAP significantly reduces the hazard for end of stay. The present analysis confirms the findings of Section 9.2.1 that were obtained from comparing the Nelson-Aalen estimates for the transitions $0 \rightarrow 2$ and $1 \rightarrow 2$.

We also note that introducing further baseline covariates into the model is technically straightforward:

```
> cox.hap.tdp <- coxph(Surv(start, stop, status) ~ pneu
+                        + age + sex, icu.pneu)
> summary(cox.hap.tdp)

  n= 1421

              coef exp(coef)  se(coef)      z Pr(>|z|)
pneu1  -0.425438  0.653484  0.106610 -3.991 6.59e-05
age     0.001355  1.001356  0.001582  0.856    0.392
sexM   -0.109530  0.896256  0.056614 -1.935    0.053

       exp(coef) exp(-coef) lower .95 upper .95
pneu1     0.6535     1.5303    0.5303    0.8053
age       1.0014     0.9986    0.9983    1.0045
sexM      0.8963     1.1158    0.8021    1.0014
```

11.2 Time-dependent covariates and competing risks

The connection between multistate models and time-dependent covariates described in Section 11.1.1 conceptually also covers the case of competing endpoints. As a consequence, a proportional hazards model with time-dependent

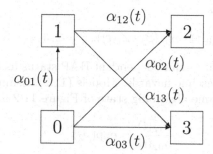

Fig. 11.2. Illness-death model without recovery and with two competing endpoints.

covariates analogous to Section 11.1.2 for the *cause-specific hazards* can immediately be formulated. In contrast, relating the *subdistribution hazard* to a time-dependent covariate requires a more detailed discussion. We consider cause-specific hazards in Section 11.2.1 and the subdistribution hazard in Section 11.2.2.

Again, things are most easily understood via an example. To be specific, we reconsider the situation of hospital-acquired pneumonia (HAP) from Section 11.1, but we now distinguish between hospital discharge (alive) and hospital death. For illustration, consider the extended illness-death model in Figure 11.2. The interpretation of the multistate model for the present example is that patients are initially in state 0. As before, occurrence of HAP is modelled by transitions into the intermediate state 1. Hospital death is modelled by transitions into state 2, and transitions into the competing endpoint state 3 reflect alive discharge. Still writing T for length of hospital stay, we now have that

$$T := \inf\{t > 0 \,|\, X_t \in \{2, 3\}\}.$$

The cumulative incidence function for hospital death is $P(T \leq t, X_T = 2)$ and the cumulative incidence function for alive discharge is $P(T \leq t, X_T = 3)$. The cause-specific hazards are

$$\alpha_j(t)\mathrm{d}t := P(T \in \mathrm{d}t, X_T = j \,|\, T \geq t), \; j = 2, 3.$$

Note that we have used a slightly different notation for the cause-specific hazards in the preceding display as compared to earlier chapters of the book, where we have indexed the cause-specific hazards by the transition type $0j$. In the present situation, we have only used the 'target state' j as an index, which reflects that the 'starting state' of the transition will also depend on the current covariate level.

11.2.1 Cause-specific hazards

A proportional cause-specific hazards model analogous to (11.1) is now immediately formulated as

$$\alpha_{j;i}(t; Z_i(t)) = \alpha_{j;0}(t) \cdot \exp\left(\beta_j \cdot Z_i(t)\right), \; j = 2, 3, \; i = 1, \ldots, n, \qquad (11.2)$$

where $Z_i(t)$ is individual i's time-dependent HAP status as before and $\alpha_{j;0}(t)$ are the cause-specific baseline hazards. Models (11.2) assume that the transition hazards into the same absorbing state of Figure 11.2 are proportional,

$$\frac{\alpha_{12}(t)}{\alpha_{02}(t)} = \exp(\beta_2)$$

$$\frac{\alpha_{13}(t)}{\alpha_{03}(t)} = \exp(\beta_3)$$

We reiterate that these models in general do not allow us to predict probabilities, because probabilities also depend on the infection hazard, the $0 \to 1$ transition.

As in Section 11.1.2, the practical key step to fit models (11.2) is to generate appropriate data sets, which may then be used with, e.g., coxph. Recall the hypothetical individual 1, who acquired HAP on day 3 with end of hospital stay on day 10. Assuming that the individual dies in hospital, the data set

contribution for analysing the cause-specific hazard of hospital death would
be:

```
id start  stop  z.t  status
1    0     4     0     0
1    4     10    1     1
```

The data set contribution for analysing the cause-specific hazard of alive discharge only differs in the last entry of status:

```
id start  stop  z.t  status
1    0     4     0     0
1    4     10    1     0
```

11.2.2 Subdistribution hazard

Incorporating time-dependent covariates in a model for the subdistribution
hazard may at first seem less straightforward as compared to modelling cause-
specific hazards. Recall from Section 5.3 that the basis of the subdistribution
framework is to consider a time ϑ until an event of interest, say hospital
death. This subdistribution time is defined to be infinite, if the competing
event occurs.

Also recall from Section 5.3 that a *technical* difficulty of the subdistribution framework is that individuals with an observed competing event should
remain in the subdistribution risk set until their potential future censoring
time; see Figure 5.12. The *conceptual* concern when relating a time-dependent
covariate to the subdistribution hazard now is the question of how the time-
dependent covariate should be defined on the interval from the real-life failure
time T to the future censoring time C if the competing event has been observed at T (Latouche et al., 2005). Following Beyersmann and Schumacher
(2008), we now illustrate that this concern disappears if we consider subdistribution hazards from a subdistribution process point of view as in Section 5.3.1.
The solution will coincide with considering a *stopped* covariate process as in
Section 11.1. In the data example, we show that the subsequent analysis provides for a synthesis of cause-specific hazards analyses as in Section 11.2.1.
We must, however, reiterate that none of these analyses allows us to predict
probabilities. We also recall that a subdistribution hazard analysis typically
assumes random censoring, whereas censoring may depend on covariates in
an analysis of the cause-specific hazards.

The idea of the subdistribution process is to stop the original process just
prior to T, if the process moves into the competing event state. Adapting
Equation (5.32) to X_t of Figure 11.2, the subdistribution process ξ_t is seen to
be

$$\xi(t) := \mathbf{1}(X(t) \neq 3) \cdot X(t) + \mathbf{1}(X(t) = 3) \cdot X(T-).$$

The subdistribution process has one absorbing state 2 and two transient
states 0 and 1. The subdistribution failure time is

$$\vartheta := \inf\{t > 0 \,|\, \xi_t = 2\}$$

with subdistribution hazard

$$\lambda(t)\mathrm{d}t := \mathrm{P}(\vartheta \in \mathrm{d}t \,|\, \vartheta \geq t).$$

The transient states of the subdistribution process are now interpreted as time-dependent covariate values as in Section 11.1.1. That is, the time-dependent covariate reflects the transitions between the transient states of ξ_t. Readers can easily verify that this amounts to considering the same time-dependent covariate

$$t \mapsto Z(t \wedge T)$$

as before!

The reason (and the solution) is that the present connection between transient states of a multistate model and time-dependent covariates implies that time-dependent covariates reflect transitions between the transient states. Because there are no further transitions after absorption, the time-dependent covariate is stopped as in the preceding display and as in Figure 11.1. And because the subdistribution process only differs from the original process if stopped at the real-life failure time, the transitions between transient states are the same both for X_t and ξ_t. Hence, we may formulate the proportional subdistribution hazards model

$$\lambda_i(t; Z_i(t)) = \lambda_0(t) \cdot \exp\left(\gamma \cdot Z_i(t)\right), \; i = 1, \ldots, n. \tag{11.3}$$

A concrete example is given below.

11.2.3 Example

Impact of hospital-acquired pneumonia (HAP) on intensive care unit mortality

We return to the HAP example of Section 11.1.3, this time distinguishing the competing events discharge alive and death in the unit. Recall from the data set description in Chapter 1 that an *observed* outcome is in `event`, but that the entry in `event` has no meaning for censored observations. Observed end of stay is indicated by `icu.pneu$status` equal to 1, which is 0 otherwise. We first create a new variable which encodes both observed competing event status and censoring status:

```
> icu.pneu$outcome <- with(icu.pneu, status * event)
```

We now fit the proportional cause-specific hazards models including HAP as the time-dependent variable.

```
> cox.hap.death <- coxph(Surv(start, stop, outcome == 2) ~ pneu,
+                        icu.pneu)
> cox.hap.disch <- coxph(Surv(start, stop, outcome == 3) ~ pneu,
+                        icu.pneu)
> summary(cox.hap.death)

  n= 1421


            coef exp(coef) se(coef)      z Pr(>|z|)
pneu1 -0.01231   0.98776  0.24668  -0.05     0.96


      exp(coef) exp(-coef) lower .95 upper .95
pneu1    0.9878      1.012    0.6091     1.602

> summary(cox.hap.disch)

  n= 1421


           coef exp(coef) se(coef)        z Pr(>|z|)
pneu1 -0.5024     0.6051   0.1185   -4.241 2.23e-05


      exp(coef) exp(-coef) lower .95 upper .95
pneu1    0.6051      1.653    0.4797    0.7632
```

Again, these findings confirm the nonparametric analyses of Section 9.2. HAP increases hospital mortality via prolonged stay, and the death hazard remains essentially unchanged.

A formal summary of these findings is achieved using subdistribution hazards. Recall from Section 5.3 that there are several ways to fit a proportional subdistribution hazards model in practice. Because the cmprsk package does not offer the possibility to consider time-dependent covariates, we use the kmi package (Allignol and Beyersmann, 2010); see Section 5.3.3. Briefly, kmi tries to recover the missing censoring times for individuals who have experienced a competing event within a multiple imputation framework. Then, e.g., the coxph function can be used on the imputed data sets to estimate subdistribution hazard ratios.

We first compute the imputed data sets using the kmi function.

```
> set.seed(4284)
> imp.dat <- kmi(Surv(start, stop, outcome != 0) ~ 1,
+                data = icu.pneu, etype = outcome,
+                id = id, failcode = 2, nimp = 10)
```

The important addition to the code of Section 5.3.3 is the id argument. It is used to identify individual subjects when one subject can have several rows of data as in our case with a time-dependent covariate. This is important because imputation of the missing censoring times is done using the observed censoring distribution. In a `coxph` framework, observations are artificially censored when the time-dependent covariate changes its value. This artificial censoring should not be used in the imputation procedure. We also specified `failcode = 2` as death is the event of interest. We now fit the Cox models on the imputed data sets using the `cox.kmi` function.

```
> kmi.sh.hap <- cox.kmi(Surv(start, stop, outcome == 2) ~ pneu,
+                          imp.dat)
> summary(kmi.sh.hap)

Call:
cox.kmi(formula = Surv(start, stop, outcome == 2) ~ pneu,
                                     imp.data = imp.dat)


****************
Pooled estimates:
****************
        coef exp(coef) se(coef)      t Pr(>|t|)
pneu1 1.108     3.028    0.240 4.616 3.91e-06


        exp(coef) exp(-coef) lower .95 upper .95
pneu1     3.028     0.3302     1.892     4.847
```

The estimated subdistribution hazard ratio is significantly larger than 1. This is a formal verification of how we interpreted the analyses of the cause-specific hazards: HAP increases intensive care unit mortality.

11.3 Further topics in the analysis of time-dependent covariates

Time-dependent bias: what happens if one treats a time-dependent exposure as being time-fixed?

In an illness-death model, the intermediate state is often interpreted as a time-dependent exposure. Being in the intermediate state means that exposure has just happened or has happened in the past. In previous examples, we have looked at hospital-acquired infection. A common mistake, sometimes called time-dependent bias (van Walraven et al., 2004), is to analyse the time-dependent exposure as if it were baseline information. In a Cox analysis, time-dependent bias would arise by coding a time-dependent exposure as a baseline covariate. For the exemplary individual of Section 11.1.2, bias arises if we substitute the correct data set entries

```
id start  stop  z.t  status
1    0     4     0     0
1    4     10    1     1
```

by the single incorrect data line

```
id start  stop  z.t  status
1    0     10    1     1
```

The single data line from the preceding display should not be used!

The bias is tantamount to ignoring that there is a $0 \to 1$ transition in the illness-death model, and that $0 \to 1$ transitions occur over the course of time. A quick calculation in terms of the Nelson-Aalen estimators shows that time-dependent bias leads to overestimation of the $0 \to 2$ hazard and to an underestimation of the $1 \to 2$ hazard (Beyersmann et al., 2008a). For a Cox model of a survival hazard, time-dependent bias will therefore artificially inflate a protective effect. A harmful effect will be artificially damped down or even reversed into an artificial protective effect (Beyersmann et al., 2008b).

An early example of time-dependent bias comes from heart transplant research (Mantel and Byar, 1974). There, a biased analysis showed a beneficial effect of transplantation on survival, but this beneficial effect was in part artificial. It owed to the fact that a transplanted patient had to stay alive on the waiting list long enough in order to finally undergo transplantation. In oncology, time-dependent bias may, e.g., arise in comparisons of patient of survival by tumor response (Anderson et al., 1983, 2008).

Readers should be aware of the fact that involved definitions of patient cohorts may quite subtly introduce time-dependent bias to the analysis. Suissa (2008), who uses the term 'immortal time bias', gives a careful discussion within the context of pharmacoepidemiology.

Plots for illustrating the effect of a time-dependent covariate

Reporting the effect of a baseline covariate such as gender in terms of hazard ratios is often supplemented by Kaplan-Meier plots, where the Kaplan-Meier curves are computed separately within the group of female patients and male patients, respectively. Similar plots for a time-dependent covariate are hampered by the fact that the model typically does not allow for predicting probabilities anymore; see our discussion following Equation (11.1). A natural choice for graphical illustration would be to present Nelson-Aalen plots for the hazards that are being related via the Cox model. For model (11.1), one might plot the Nelson-Aalen estimators of $\int_0^t \alpha_{02}(u)\mathrm{d}u$ and of $\int_0^t \alpha_{12}(u)\mathrm{d}u$. The advantage of such plots is that they closely relate to the model used for the analysis. The inevitable disadvantage is that such plots do not depict probabilities.

An alternative is the so-called 'landmark method' (e.g., Anderson et al., 1983). The idea is to choose a number of landmarks s, $s > 0$. The landmark s serves as a new time origin and patients are grouped according to their covariate value at time s. Then, Kaplan-Meier curves may be calculated within groups and starting at time s. The advantage of landmarking is that it produces probability plots, which may be easier to communicate. The disadvantage is that landmarking produces an increased number of plots, and that the depicted probabilities do not directly relate to the model used for the analysis, but also depend on the development of the time-dependent covariate. We note that there is renewed statistical interest in landmarking (e.g., van Houwelingen and Putter, 2008). Landmark plots have been produced in Section 9.2.1.

Sometimes, efforts are made to obtain other Kaplan-Meier-type plots in the presence of time-dependent covariates. Some arise from time-dependent bias as discussed above and rely on retrospectively grouping patients by their final covariate value. Others arise from computing a Kaplan-Meier-type statistic based on the Nelson-Aalen estimator of one transition hazard only. These plots do not have a meaningful probability interpretation and should not be used.

Should one adjust for a time-dependent covariate when comparing baseline treatment groups?

In the analyses of hospital-acquired pneumonia considered earlier, there was explicit research interest in investigating the impact of the infection. A related but also somewhat different issue arises, e.g., in clinical trials where the main aim is to compare treatment groups. Often, treatment is decided upon at time zero, e.g., via randomization, and patients are prospectively followed until the occurrence of an event or closure of the study. Time-dependent covariates may be collected during follow-up. The question now is whether to include the time-dependent covariates in, e.g., a Cox model, when the aim is to compare the treatments. The concern is that time-dependent covariates can be 'responsive' (Kalbfleisch and Prentice, 2002, p. 199). An example is tumor response to treatment (e.g, Anderson et al., 2008). Assume that a beneficial treatment effect is mediated via changes in the time-dependent covariate. Then, including the time-dependent covariate in the model may give insights into this mechanism, but it may also diminish the beneficial treatment effect as displayed in the analysis. Therefore, it is typically recommended to leave out the time-dependent covariate from the analysis. This issue is obviously subtle, but relevant, and it is still a topic of current statistical research. We refer to Chapters 8 and 9 of Aalen et al. (2008) for very useful textbook accounts.

Investigating and relaxing the Markov assumption

The most common departures from the Markov property are duration dependencies or dependency on the number of previous recoveries, say, in an illness-death model with recovery. In particular, the Markov assumption in the important illness-death model without recovery is tantamount to assuming that the $1 \to 2$ hazard only depends on time t since time origin, but not on the entry time, say s, into the intermediate state 1. This suggests ways to both investigate and relax the Markov assumption via time-dependent covariates.

To be specific, consider the illness-death model *without* recovery. Then we may investigate the Markov assumption by including the entry time s into state 1 (e.g., Keiding and Gill, 1990) or the waiting time $t - s$ in state 1 (e.g., Andersen and Keiding, 2002) in, e.g., a Cox model for the $1 \to 2$ hazard, the hazard of 'dying' after 'illness'. If there is evidence for a departure from the Markov property, one may choose to model this by keeping the waiting time in state 1 in a regression model for the $1 \to 2$ hazard. In more complex models such as the illness-death model with recovery, one may choose to include the number of previous transitions into the 'illness'-state into the model for the $0 \to 1$ hazard of getting 'ill', or one may consider a time-dependent indicator of whether one has been ill before.

We note that Allignol et al. (2011b) investigated the Markov assumption for the hospital-acquired pneumonia data used earlier in this book. The authors found it reasonable to assume the Markov property.

Investigating the assumption of independent left-truncation

An idea analogous to investigating the Markov assumption applies when the aim is to assess the assumption of independent or random left-truncation. To be specific, consider a standard survival situation, where data are subject to left-truncation and potentially also right-censoring. Then, the assumption of independent or random left-truncation may be investigated by including the study entry time into a Cox model for the all-cause hazard. This analogy to investigating the Markov assumption is not a coincidence: Keiding and Gill (1990) gave a fundamental study of left-truncation by using a reparametrization in terms of a Markovian multistate model.

We note that Allignol et al. (2010) assessed the assumption of random left-truncation for the pregnancy outcome data that were analysed in Sections 4.4 and 5.2.2. The authors argued that random left-truncation may reasonably be assumed.

11.4 Exercises

1. Redo the proportional subdistribution hazards analysis of Section 11.2.2 using the administrative censoring times contained in `adm.cens.exit`. (Hint: See Section 5.3.3 on using adminstrative censoring times.)
2. Consider the ventilation data analysed in, e.g., Section 10.2.2.
   a) Fit a Cox proportional hazards model for the hazard of discharge, considering ventilation status as a time-dependent covariate.
   b) Fit the same model, but with an additional binary time-dependent covariate indicating prior ventilation.

3. Reconsider Exercise 1 of Section 10.3. Redo the analyses, including as a time-dependent covariate the information whether one has recovered from 'illness' in the past.

4. *Time-dependent bias*: Show that an analysis subject to time-dependent bias underestimates the cumulative hazard of reaching the absorbing state for exposed individuals. Also show that the biased analysis overestimates the corresponding cumulative hazard for non-exposed individuals. (Hint: Consider the Nelson-Aalen estimators for an illness-death model without recovery.)

12

Further topics in multistate modelling

More parsimonious Cox-type regression models

In Section 5.2 on proportional cause-specific hazards models, we illustrated that one typically has to investigate the effect of a covariate on all transition hazards. We also outlined that some covariates may have a common effect on some of the cause-specific hazards, but this is rarely used in practical applications. In more complex multistate models, however, sample size restrictions may motivate more parsimonious models. In practice, this is typically achieved by analysing an extended data frame as in Section 5.2.2 (see the data frame x1) with one row for each individual and each transition. Such an extended data frame for multistate data has been discussed in Section 10.2. As illustrated in Section 5.2.2, transition-specific covariates are used which allow single covariates to have a common effect on some transition hazards and different effects on other hazards. One may also impose that a covariate has no effect on a certain hazard by setting the corresponding entry of the transition-specific covariate to zero. Finally, one may assume some baseline transition hazards to be proportional, say, $\alpha_{lj;0}(t) = \exp(\beta) \cdot \alpha_{\tilde{l}\tilde{j};0}(t)$. This is achieved by introducing an additional dummy column to the extended data frame with entry 1 for the $\tilde{l} \rightarrow \tilde{j}$ transition and 0 otherwise.

While these techniques are useful, more practical experience for model building is needed. A concise description of the techniques at hand is also provided by Andersen and Keiding (2002) and de Wreede et al. (2010).

Investigating and relaxing the Markov assumption

See Section 11.3.

Beyond the Markov assumption

In Section 11.3, we have explained that one may model departures from the Markov property by including entry times into a state, waiting times in a state, or the number of previous visits to a state in Cox-type regression models for the transition hazards. We also discussed how such approaches can be used to investigate whether the Markov property holds. In this paragraph, we comment on some approaches to nonparametric estimation for non-Markov multistate models.

If the data are subject to random censoring only (i.e., censoring is completely independent of the multistate process) and if the process starts in one common initial state 0, Datta and Satten (2001) showed that the Aalen-Johansen estimator of $P(X_t = j \mid X_0 = 0)$ is a consistent estimator of the state occupation probability $P(X_t = j)$ even in the absence of the Markov property. Glidden (2002) provides weak convergence results towards a Gaussian process. However, variances are even more complicated than in the Markovian case. Therefore, it is more convenient to use the bootstrap as in Appendix A to estimate variances.

If the initial distribution of the process is not degenerated, one may estimate the initial distribution by the observed relative frequencies in the states of the model at time 0 and combine these estimates with the Aalen-Johansen estimators of $P(X_t = j \mid X_0 = l)$ in the usual way to obtain an estimator of the state occupation probability $P(X_t = j)$.

Models, which are only forward moving, may typically be reparametrized by multivariate event times. The simple illness-death model *without* recovery and common initial state 0 can equivalently be expressed by the pair of event times (e.g., Beyersmann, 2007)

$$T_0 := \inf\{t \ge 0 \mid X_t \ne 0\} \text{ and } T := \inf\{t \ge 0 \mid X_t = 2\}.$$

In this parametrization, we have $T_0 \le T$, and $T - T_0$ is the length of time spent in the intermediate state 1. For direct $0 \to 1$ transitions, we have $T - T_0 = 0$. The idea is to estimate the joint distribution of (T_0, T). Then, transition probabilities may be derived as deterministic functions of the joint distribution. E.g., we have that $P(X_t = 2 \mid X_s = 1)$ is equal to $P(T \le t \mid T_0 \le s < T)$.

The joint distribution of (T_0, T) in the presence of random right-censoring may be estimated by techniques of multivariate survival analysis. However, multivariate survival analysis in general poses challenging problems. One issue is that the concepts of 'past' and 'future' are not unambiguous anymore in multivariate time (Gill, 1992). Fortunately, these problems reduce to a certain degree in the present context, where (T_0, T) are subject to a common censoring variable (Tsai and Crowley, 1998).

Meira-Machado et al. (2006) have used the idea to estimate transition probabilities in a non-Markov illness-death model without recovery. Their estimator is implemented in the R package `p3state.msm` (Meira-Machado and Pardinas, 2011).

Choice of time origin

Typically, the choice of time origin and, hence, the time scale is governed by subject matter considerations. In randomized clinical trials, a natural choice for time 0 is time of randomization. In some epidemiologic studies, age may be the time scale of choice. We believe that time is conceptually so fundamental that, if possible, it is best not to change time scales in order to avoid confusion.

However, there may be situations where one may consider different time scales. In a non-Markov illness-death model without recovery and common initial state 0, the $1 \to 2$ hazard will depend on time t since time origin and on time s of entry into state 1. A semi-Markov or Markov renewal process arises, if the hazard only depends on the duration $t - s$. In such a situation, one may consider to choose the time of state entry as the new time origin. Putter et al. (2007) refer to this as 'clock reset', whereas keeping the origin time scale is referred to as 'clock forward'. Andersen and Keiding (2002) argue that the 'clock reset' approach is useful in Cox-type regression modelling of the $1 \to 2$ hazard, if the effect of the entry time into the intermediate state is that 'irregular' such that it is best captured via a nonparametric baseline hazard rather than by parametrically modelling it via a time-dependent covariate.

Homogeneous Markov processes

This book has focused on the usual non- and semiparametric techniques for analysing competing risks and multistate data subject to independent right-censoring and left-truncation. These methods are especially common in biostatistical applications, where it is often difficult to justify a particular parametric model. In fact, in Section 8.1, we emphasized that the multistate models at hand were time-inhomogeneous Markov processes with time-varying transition hazards rather than homogeneous processes with time-constant hazards. But there will, of course, be situations where working with (piecewise) constant hazards is attractive.

One such situation is large data sets. The estimator of the $l \to j$ hazard under a constant hazard assumption is the number of all observed $l \to j$ transitions divided by the total time at risk in state l. For variance estimation, one divides by the square of the total time at risk in state l. In other words, the only information that one needs to store is the number of all observed $l \to j$ transitions and the total time at risk in state l, aggregated over all individuals, but the individual information is not needed for such an analysis. This is attractive when handling truly large data sets.

Another situation is interval-censored data. Interval censoring arises if the occurrence of an event is only known to fall in some time interval, but the exact event time is not known. Accounting for interval censoring in non- and semiparametric inference is challenging, but this observational pattern is easily incorporated when assuming constant hazards.

In R, multistate models with (piecewise) constant hazards are typically analysed using the R package `msm`. Jackson (2011) gives an excellent account of the package, along with a concise and clear description of data situations, where the approach will be useful.

Investigating independent left-truncation

See Section 11.3.

Nonparametric hypothesis testing

In Chapter 6, we have explained that the log-rank test may be used to compare the cause-specific hazards of a type 1 event, say, between groups. In practice, these tests may be computed as a byproduct when fitting Cox models or using the function `survdiff` of the package `survival`. Analogously, a log-rank test may be computed to compare the $l \to j$ transition hazard between groups for fixed $l, j, l \neq j$.

It also holds analogously to competing risks that testing equality of the $l \to j$ transition hazard is not tantamount to testing equality of a certain transition probability. Little work has been done if the aim is to compare transition probabilities. One approach would be to directly model the probabilities as in Andersen et al. (2003) or in Scheike and Zhang (2007); see also Section 5.6. Another approach would be to base tests on confidence *bands* for the group difference of the transition probability under consideration. This approach will typically build on some resampling procedure as in Appendix A to construct the confidence bands. Lin (1997) discusses one approach for competing risks.

Direct regression models for the transition probabilities

In the competing risks part of the book, we discussed Cox-type regression models for both the cause-specific hazards and the subdistribution hazard. The analysis of the subdistribution hazard allowed for a direct interpretation of the effect of a covariate on one cumulative incidence function, and analysing the cause-specific hazards was required to understand how such an effect was mediated. In the multistate models part of the book, we considered Cox-type regression models for the transition hazards in obvious analogy to the cause-specific hazards. Direct regression models for the transition probabilities of a general multistate model have, e.g., been developed by Andersen et al. (2003) and Scheike and Zhang (2007). We have briefly discussed their approaches in Section 5.6. A very useful review paper is Andersen and Perme (2008).

A

A nonparametric bootstrap for multistate models

Briefly speaking, the bootstrap (Efron, 1979) is a computer resampling experiment, which samples according to the empirical distribution of the data. The idea is that the computer experiment should 'work', if the empirical distribution is a reasonable approximation of the true underlying distribution. One typical application of the bootstrap is to estimate variances, if a variance estimator is algebraically complicated or if a formula for variance estimation is not available at all. Andersen et al. (1993) (p. 221) value the bootstrap as 'an attractive alternative to the calculation of a complicated asymptotic distribution.'

For multistate process data, there are different bootstrap approaches; see, e.g., the brief discussion in Section IV.1.4 of Andersen et al. (1993). Here, we describe Efron's approach (Efron, 1981), originally introduced for i.i.d. randomly censored survival data. Given that there are n individuals under study, the resampling consists of drawing n times with replacement from the original individuals. Such new bootstrap data sets are created a large number of times, say, m times. Often, m is 1000 or even 10000.

For each of the m bootstrap data sets, the statistic of interest is computed. The statistic of interest could be the Aalen-Johansen estimator of a transition probability or the maximum likelihood estimator of a regression coefficient. The variance of the statistic of interest is then estimated as the empirical variance of the m replicates of the statistic of interest.

In Section 5.4, we used the bootstrap to estimate the variance of the estimated least false parameter. In Section 10.2.2, we used the bootstrap to estimate variances for the estimated regression coefficients when fitting Cox models for transition hazards of an illness-death model with recovery. The practical steps are illustrated below. Correctness of the bootstrap is investigated by, e.g., Gill (1989) and van der Vaart and Wellner (1996).

The following code was used for computing the bootstrapped variance estimates in Section 10.2.2. For each iteration of the loop, individuals are drawn with replacement using the `sample` function. As information for one individual can be spread across several lines, it is the id numbers that are actually sampled. As in Section 10.2.2, id is the variable which identifies the lines belonging to one person.

We then create the bootstrap data set dboot. Care must be taken here, because most matching procedures in R only select the first match. Hence, we used the more involved construct below that returns for each element of index the line numbers where the sampled id is equal to the individual id. Then the statistic of interest is computed for each of the bootstrap data sets. At the end of the nboot = 10000 bootstrap iterations, the empirical variance of the bootstrap regression coefficients is computed.

```
> nboot <- 10000
> res <- lapply(seq_len(nboot), function(i) {
+       index <- sample(unique(sir.cont$id), replace=TRUE)
+       dboot <- sir.cont[unlist(sapply(index, function(x)
+                                       which(x==sir.cont[["id"]]
+                                       ))), ]
+       coef01 <- coef(coxph(Surv(time, to == 1) ~ age + sex,
+                            dboot, subset = from == 0))
+       coef10 <- coef(coxph(Surv(time, to == 0) ~ age + sex,
+                            dboot, subset = from == 1))
+       matrix(c(coef01, coef10), ncol = 4)
+ })
> res <- do.call(rbind, res)
> se.boot <- matrix(sqrt(apply(res, 2, var)), ncol = 4,
+                   dimnames = list("", c("age01", "sex01",
+                   "age10", "sex10")))
```

We also mention the msboot function in the mstate package. The function randomly samples with replacement subjects from the original dataset, provided that the original data set is in the msdata format.

References

Aalen, O. (1980). A model for nonparametric regression analysis of counting processes. *Springer Lecture Notes in Statistics*, 2:1–25, Springer, New York.

Aalen, O. (1987). Dynamic modelling and causality. *Scandinavian Actuarial Journal*, pages 177–190.

Aalen, O., Borgan, Ø., and Fekjær, H. (2001). Covariate adjustment of event histories estimated from Markov chains: the additive approach. *Biometrics*, 57:993–1001.

Aalen, O., Borgan, Ø., and Gjessing, H. (2008). *Event History Analysis*. Springer, New York.

Aalen, O. and Johansen, S. (1978). An empirical transition matrix for non-homogeneous Markov chains based on censored observations. *Scandinavian Journal of Statistics*, 5:141–150.

Allignol, A. and Beyersmann, J. (2010). Software for fitting non-standard subdistribution hazards models. *Biostatistics*, 11(4):674–675.

Allignol, A., Beyersmann, J., and Schumacher, M. (2008). mvna: An R package for the Nelson-Aalen estimator in multistate models. *R News*, 8(2):48–50.

Allignol, A., Schumacher, M., and Beyersmann, J. (2010). A note on variance estimation of the Aalen-Johansen estimator of the cumulative incidence function in competing risks, with a view towards left-truncated data. *Biometrical Journal*, 52(1):126–137.

Allignol, A., Schumacher, M., and Beyersmann, J. (2011a). Empirical transition matrix of multistate models: the etm package. *Journal of Statistical Software*, 38(4):1–15.

Allignol, A., Schumacher, M., and Beyersmann, J. (2011b). Estimating summary functionals in multistate models with an application to hospital infection data. *Computational Statistics*, 26:181–197.

Allignol, A., Schumacher, M., Wanner, C., Drechsler, C., and Beyersmann, J. (2011c). Understanding competing risks: a simulation point of view. *BMC Medical Research Methodology*. To appear.

Andersen, P. (1986). Time-dependent covariates and Markov processes. In Moolgavkar, S. and Prentice, R., editors, *Modern Statistical Methods in Chronic Disease Epidemiology*, pages 82–103. Wiley, New York.

Andersen, P., Abildstrøm, S., and Rosthøj, S. (2002). Competing risks as a multistate model. *Statistical Methods in Medical Research*, 11(2):203–215.

Andersen, P. and Borgan, Ø. (1985). Counting process models for life history data: A review. *Scandinavian Journal of Statistics*, 12:97–140.

Andersen, P., Borgan, Ø., Gill, R., and Keiding, N. (1993). *Statistical Models Based on Counting Processes*. Springer, New York.

Andersen, P. and Gill, R. (1982). Cox's regression model for counting processes: A large sample study. *The Annals of Statistics*, 10:1100–1120.

Andersen, P. and Keiding, N. (2002). Multi-state models for event history analysis. *Statistical Methods in Medical Research*, 11(2):91–115.

Andersen, P., Klein, J., and Rosthøj, S. (2003). Generalised linear models for correlated pseudo-observations with applications to multi-state models. *Biometrika*, 90(1):15–27.

Andersen, P. and Perme, M. (2008). Inference for outcome probabilities in multistate models. *Lifetime Data Analysis*, 14(4):405–431.

Andersen, P. and Perme, M. (2010). Pseudo-observations in survival analysis. *Statistical Methods in Medical Research*, 19(1):71–99.

Anderson, J., Cain, K., and Gelber, R. (1983). Analysis of survival by tumor response. *Journal of Clinical Oncology*, 1:710–719.

Anderson, J., Cain, K., and Gelber, R. (2008). Analysis of survival by tumor response and other comparisons of time-to-event by outcome variables. *Journal of Clinical Oncology*, 26(24):3913–3915.

Bajorunaite, R. and Klein, J. (2008). Comparison of failure probabilities in the presence of competing risks. *Journal of Statistical Computation and Simulation*, 78(10):951–966.

Barnett, A., Batra, R., Graves, N., Edgeworth, J., Robotham, J., and Cooper, B. (2009). Using a longitudinal model to estimate the effect of methicillin-resistant staphylococcus aureus infection on length of stay in an intensive care unit. *American Journal of Epidemiology*, 170(9):1186–1194.

Bender, R., Augustin, T., and Blettner, M. (2005). Generating survival times to simulate Cox proportional hazards models. *Statistics in Medicine*, 24:1713–1723.

Bernoulli, D. (1982). *Die gesammelten Werke der Mathematiker und Physiker der Familie Bernoulli. Hrsg. von der Naturforschenden Gesellschaft in Basel. Die Werke von Daniel Bernoulli. Band 2: Analysis, Wahrscheinlichkeitsrechnung.* Birkhäuser, Basel.

Bernoulli, D. and Blower, S. (2004). An attempt at a new analysis of the mortality caused by smallpox and of the advantages of inoculation to prevent it. *Reviews in Medical Virology*, 14(5):275–288.

Beyersmann, J. (2007). A random time interval approach for analysing the impact of a possible intermediate event on a terminal event. *Biometrical Journal*, 49(5):742–749.

Beyersmann, J., Dettenkofer, M., Bertz, H., and Schumacher, M. (2007). A competing risks analysis of bloodstream infection after stem-cell transplantation using subdistribution hazards and cause-specific hazards. *Statistics in Medicine*, 26(30):5360–5369.

Beyersmann, J., Gastmeier, P., Wolkewitz, M., and Schumacher, M. (2008a). An easy mathematical proof showed that time-dependent bias inevitably leads to biased effect estimation. *Journal of Clinical Epidemiology*, 61(12):1216–1221.

Beyersmann, J., Latouche, A., Buchholz, A., and Schumacher, M. (2009). Simulating competing risks data in survival analysis. *Statistics in Medicine*, 28:956–971.

Beyersmann, J. and Schumacher, M. (2008). Time-dependent covariates in the proportional subdistribution hazards model for competing risks. *Biostatistics*, 9:765–776.

Beyersmann, J., Wolkewitz, M., Allignol, A., Grambauer, N., and Schumacher, M. (2011). Application of multistate models in hospital epidemiology: advances and challenges. *Biometrical Journal*, 53:332–350.

Beyersmann, J., Wolkewitz, M., and Schumacher, M. (2008b). The impact of time-dependent bias in proportional hazards modelling. *Statistics in Medicine*, 27:6439–6454.

Bie, O., Borgan, Ø., and Liestøl, K. (1987). Confidence intervals and confidence bands for the cumulative hazard rate function and their small sample properties. *Scandinavian Journal of Statistics*, 14:221–233.

Billingsley, P. (1968). *Convergence of Probability Measures*. Wiley, New York.

Bradley, L. (1971). *Smallpox Inoculation: An Eighteenth Century Mathematical Controversy*. Adult Education Department of the University of Nottingham.

Braun, T. and Yuan, Z. (2007). Comparing the small sample performance of several variance estimators under competing risks. *Statistics in Medicine*, 26(5):1170–1180.

Breslow, N. (1972). Discussion of the paper by D.R.Cox. *Journal of the Royal Statistical Society, Series B*, 34:216–217.

Burton, A., Altman, D., Royston, P., and Holder, R. (2006). The design of simulation studies in medical statistics. *Statistics in Medicine*, 25:4279–4292.

Cheng, S., Fine, J., and Wei, L. (1998). Prediction of cumulative incidence function under the proportional hazards model. *Biometrics*, 54:219–228.

Claeskens, G. and Hjort, N. (2008). *Model Selection and Model Averaging*. Cambridge University Press, Cambridge.

Clark, T., Altman, D., and De Stavola, B. (2002). Quantification of the completeness of follow-up. *The Lancet*, 359:1309–1310.

Cortese, G. and Andersen, P. (2010). Competing risks and time-dependent covariates. *Biometrical Journal*, 52(1):138–158.

Cox, D. (1972). Regression models and life-tables. *Journal of the Royal Statistical Society, Series B: Methodological*, 34:187–220.

D'Agostino, R., Lee, M., Belanger, A., Cupples, L., Anderson, K., and Kannel, W. (1990). Relation of pooled logistic regression to time dependent Cox regression analysis: the Framingham Heart Study. *Statistics in Medicine*, 9(12):1501–1515.

Dalgaard, P. (2002). *Introductory statistics with R*. Springer, New York.

Datta, S. and Satten, G. A. (2001). Validity of the Aalen-Johansen estimators of stage occupation probabilities and Nelson-Aalen estimators of integrated transition hazards for non-Markov models. *Statistics and Probability Letters*, 55(4):403–411.

de Wreede, L., Fiocco, M., and Putter, H. (2010). The mstate package for estimation and prediction in non- and semi-parametric multi-state and competing risks models. *Computer Methods and Programs in Biomedicine*, 99:261–274.

de Wreede, L. C., Fiocco, M., and Putter, H. (2011). mstate: An R package for the analysis of competing risks and multi-state models. *Journal of Statistical Software*, 38(7):1–30.

Dettenkofer, M., Wenzler-Röttele, S., Babikir, R., Bertz, H., Ebner, W., Meyer, E., Rüden, H., Gastmeier, P., and Daschner, F. (2005). Surveillance of nosocomial sepsis and pneumonia in patients with a bone marrow or peripheral blood stem cell transplant. A multicenter project. *Clinical Infectious Diseases*, 40:926–931.

Dietz, K. and Heesterbeek, J. (2002). Daniel Bernoulli's epidemiological model revisited. *Mathematical Biosciences*, 180:1–21.

Efron, B. (1979). Bootstrap methods: Another look at the jackknife. *Annals of Statistics*, 7:1–26.

Efron, B. (1981). Censored data and the bootstrap. *Journal of the American Statistical Association*, 76:312–319.

Fine, J. (2002). Comparing nonnested Cox models. *Biometrika*, 89(3):635–648.

Fine, J. and Gray, R. (1999). A proportional hazards model for the subdistribution of a competing risk. *Journal of the American Statistical Association*, 94(446):496–509.

Gail, M. (1982). Competing risks. In Banks, D., Read, C., and Kotz, S., editors, *Encyclopedia of Statistical Sciences, Volume 2*, pages 75–81. Wiley, New York.

Gerds, T., Cai, T., and Schumacher, M. (2008). The performance of risk prediction models. *Biometrical Journal*, 50:457–479.

Gerds, T. and Schumacher, M. (2001). On functional misspecification of covariates in the Cox regression model. *Biometrika*, 88(2):572–580.

Geskus, R. (2011). Cause-specific cumulative incidence estimation and the Fine and Gray model under both left truncation and right censoring. *Biometrics*, 67:39–49.

Gill, R. (1984). Understanding Cox's regression model: A martingale approach. *Journal of the American Statistical Association*, 79:441–447.

Gill, R. (1989). Non- and semi-parametric maximum likelihood estimators and the von Mises method (Part 1). *Scandinavian Journal of Statistics*, 16(2):97–128.

Gill, R. (1992). Multivariate survival analysis. *Theory of Probability and its Applications*, 37(1):18–31.

Gill, R. and Johansen, S. (1990). A survey of product-integration with a view towards application in survival analysis. *Annals of Statistics*, 18(4):1501–1555.

Gill, R. and Schumacher, M. (1987). A simple test of the proportional hazards assumption. *Biometrika*, 74:289–300.

Glidden, D. (2002). Robust inference for event probabilities with non-Markov data. *Biometrics*, 58:361–368.

Glynn, R., Rosner, B., and Christen, W. (2009). Evaluation of risk factors for cataract types in a competing risks framework. *Ophthalmic Epidemiology*, 16(2):98–106.

Grambauer, N., Schumacher, M., and Beyersmann, J. (2010a). Proportional subdistribution hazards modeling offers a summary analysis, even if misspecified. *Statistics in Medicine*, 29:875–884.

Grambauer, N., Schumacher, M., Dettenkofer, M., and Beyersmann, J. (2010b). Incidence densities in a competing events analysis. *American Journal of Epidemiology*, 172(9):1077–1084.

Graw, F., Gerds, T., and Schumacher, M. (2009). On pseudo-values for regression analysis in competing risks models. *Lifetime Data Analysis*, 15:241–255.

Gray, R. (1988). A class of k-sample tests for comparing the cumulative incidence of a competing risk. *Annals of Statistics*, 16(3):1141–1154.

Hjort, N. (1992). On inference in parametric survival data models. *International Statistical Review*, 60:355–387.

Hosmer, D., Lemeshow, S., and May, S. (2008). *Applied Survival Analysis. Regression Modeling of Time to Event Data. 2nd ed.* Wiley, Hoboken, NJ.

Hougaard, P. (2000). *Analysis of Multivariate Survival Data.* Springer, New York.

Jackson, C. (2011). Multi-state models for panel data: the msm package for R. *Journal of Statistical Software*, 38(8):1–29.

Kalbfleisch, J. and Prentice, R. (1980). *The Statistical Analysis of Failure Time Data.* Wiley, New York.

Kalbfleisch, J. and Prentice, R. (2002). *The Statistical Analysis of Failure Time Data. 2nd ed.* Wiley, Hoboken, NJ.

Kay, R. (1982). The analysis of transition times in multistate stochastic processes using proportional hazard regression models. *Communications in Statistics-Theory and Methods*, 11(15):1743–1756.

Keiding, N. and Gill, R. (1990). Random truncation models and Markov processes. *The Annals of Statistics*, 18:582–602.

Klein, J. (1991). Small sample moments of some estimators of the variance of the Kaplan-Meier and Nelson-Aalen estimators. *Scandinavian Journal of Statistics*, 18:333–340.

Klein, J. (2006). Modelling competing risks in cancer studies. *Statistics in Medicine*, 25:1015–1034.

Klein, J., Gerster, M., Andersen, P., Tarima, S., and Perme, M. (2008). SAS and R functions to compute pseudo-values for censored data regression. *Computer Methods and Programs in Biomedicine*, 89(3):289–300.

Klein, J., Keiding, N., Shu, Y., Szydlo, R., and Goldman, J. (2000). Summary curves for patients transplanted for chronic myeloid leukaemia salvaged by a donor lymphocyte infusion: the current leukaemia-free survival curve. *British Journal of Haematology*, 20:1871–1885.

Klein, J. and Moeschberger, M. (2003). *Survival Analysis. Techniques for Censored and Truncated Data. 2nd ed.* Springer, New York.

Koller, M., Stijnen, T., Steyerberg, E., and Lubsen, J. (2008). Meta-analyses of chronic disease trials with competing causes of death may yield biased odds ratios. *Journal of Clinical Epidemiology*, 61(4):365–372.

Latouche, A. (2004). *Modèle de Régression en Présence de Compétition.* PhD thesis, Université Paris 6. http://tel.archives-ouvertes.fr/tel-00129238/fr/.

Latouche, A., Allignol, A., Beyersmann, J., Labopin, M., and Fine, J. (2011). Analysing, interpreting and reporting competing risks in stem cell transplantation: a case study in the RIC era. *Unpublished manuscript*.

Latouche, A., Boisson, V., Porcher, R., and Chevret, S. (2007). Misspecified regression model for the subdistribution hazard of a competing risk. *Statistics in Medicine*, 26(5):965–974.

Latouche, A. and Porcher, R. (2007). Sample size calculations in the presence of competing risks. *Statistics in Medicine*, 26(30):5370–5380.

Latouche, A., Porcher, R., and Chevret, S. (2005). A note on including time-dependent covariate in regression model for competing risks data. *Biometrical Journal*, 47(6):807–814.

Leisch, F. (2002). Sweave: Dynamic generation of statistical reports using literate data analysis. In Härdle, W. and Rönz, B., editors, *Compstat 2002 — Proceedings in Computational Statistics*, pages 575–580. Physica Verlag, Heidelberg.

Lin, D. (1997). Non-parametric inference for cumulative incidence functions in competing risks studies. *Statistics in Medicine*, 16:901–910.

Lin, D. and Wei, L. (1989). The robust inference for the Cox proportional hazards model. *Journal of the American Statistical Association*, 84:1074–1078.

Lin, D. and Ying, Z. (1994). Semiparametric analysis of the additive risk model. *Biometrika*, 81:61–71.

Lumley, T. (2004). The survival package. *R News*, 4(1):26–28.

Lunn, M. and McNeil, D. (1995). Applying Cox regression to competing risks. *Biometrics*, 51:524–532.

Mackenbach, J., Kunst, A., Lautenbach, H., Oei, Y., and Bijlsma, F. (1999). Gains in life expectancy after elimination of major causes of death: revised estimates taking into account the effect of competing causes. *Journal of Epidemiology and Community Health*, 53(1):32–37.

Mantel, N. and Byar, D. (1974). Evaluation of response-time data involving transient states: An illustration using heart-transplant data. *Journal of the American Statistical Association*, 69:81–86.

Martinussen, T. and Scheike, T. (2006). *Dynamic Regression Models for Survival Data*. Springer, New York.

McKeague, I. and Sasieni, P. (1994). A partly parametric additive risk model. *Biometrika*, 81:501–514.

Meira-Machado, L., de Uña-Álvarez, J., and Cadarso-Suárez, C. (2006). Nonparametric estimation of transition probabilities in a non-Markov illness-death model. *Lifetime Data Analysis*, 12(3):325–344.

Meira-Machado, L. and Pardinas, J. (2011). p3state.msm: Analyzing survival data from an illness-death model. *Journal of Statistical Software*, 38(3):1–18.

Meister, R. and Schaefer, C. (2008). Statistical methods for estimating the probability of spontaneous abortion in observational studies–analyzing pregnancies exposed to coumarin derivatives. *Reproductive Toxicology*, 26(1):31–35.

Meyer, E., Beyersmann, J., Bertz, H., Wenzler-Röttele, S., Babikir, R., Schumacher, M., Daschner, F., Rüden, H., Dettenkofer, M., and the ONKO-KISS study group (2007). Risk factor analysis of blood stream infection and pneumonia in neutropenic patients after peripheral blood stem-cell transplantation. *Bone Marrow Transplant*, 39(3):173–178.

Morgan, B. (1984). *Elements of Simulation*. Chapman & Hall, London.

Pepe, M. and Mori, M. (1993). Kaplan-Meier, marginal or conditional probability curves in summarizing competing risks failure time data? *Statistics in Medicine*, 12:737–751.

Prentice, R., Kalbfleisch, J., Peterson, A., Flournoy, N., Farewell, V., and Breslow, N. (1978). The analysis of failure times in the presence of competing risks. *Biometrics*, 34:541–554.

Putter, H., Fiocco, M., and Geskus, R. (2007). Tutorial in biostatistics: competing risks and multi-state models. *Statistics in Medicine*, 26(11):2277–2432.

Ripley, B. (1987). *Stochastic Simulation*. Wiley, New York.

Rizzo, M. (2007). *Statistical Computing with R*. Chapman & Hall, Boca Raton, FL.

Robins, J. and Rotnitzky, A. (1992). Recovery of information and adjustment for dependent censoring using surrogate markers. In Jewell, N., Dietz, K., and Farewell, V., editors, *AIDS Epidemiology — Methodological Issues*, pages 24–33. Birkhäuser, Boston.

Rosthøj, S., Andersen, P., and Abildstrom, S. (2004). SAS macros for estimation of the cumulative incidence functions based on a Cox regression model for competing risks survival data. *Computer Methods and Programs in Biomedicine*, 74:69–75.

Ruan, P. and Gray, R. (2008). Analyses of cumulative incidence functions via non-parametric multiple imputation. *Statistics in Medicine*, 27(27):5709–5724.

Russell, B. (2004). *History of Western Philosophy. Reprint of the 1946 ed.* Routledge, Oxon.

Saha, P. and Heagerty, P. (2010). Time-dependent predictive accuracy in the presence of competing risks. *Biometrics*, 66:999–1011.

Schafer, J. (1997). *Analysis of Incomplete Multivariate Data.* Chapman & Hall, London.

Schaubel, D. and Wei, G. (2007). Fitting the additive hazards model using standard statistical software. *Biometrical Journal*, 49:719–730.

Scheike, T. and Zhang, M.-J. (2007). Direct modelling of regression effects for transition probabilities in multistate models. *Scandinavian Journal of Statistics*, 34(1):17–32.

Scheike, T. and Zhang, M.-J. (2008). Flexible competing risks regression modeling and goodness-of-fit. *Lifetime Data Analysis*, 14(4):464–483.

Scheike, T., Zhang, M.-J., and Gerds, T. (2008). Predicting cumulative incidence probability by direct binomial regression. *Biometrika*, 95(1):205–220.

Schoop, R., Beyersmann, J., Schumacher, M., and Binder, H. (2011). Quantifying the predictive accuracy of time-to-event models in the presence of competing risks. *Biometrical Journal*, 53:88–112.

Schulgen, G., Olschewski, M., Krane, V., Wanner, C., Ruf, G., and Schumacher, M. (2005). Sample sizes for clinical trials with time-to-event endpoints and competing risks. *Contemporary Clinical Trials*, 26:386–395.

Schumacher, M., Wangler, M., Wolkewitz, M., and Beyersmann, J. (2007). Attributable mortality due to nosocomial infections: a simple and useful application of multistate models. *Methods of Information in Medicine*, 46:595–600.

Shiryaev, A. (1995). *Probability. 2nd ed.* Springer, New York.

Struthers, C. and Kalbfleisch, J. (1986). Misspecified proportional hazard models. *Biometrika*, 73:363–369.

Suissa, S. (2008). Immortal time bias in pharmacoepidemiology. *American Journal of Epidemiology*, 167(4):492–499.

Temkin, N. (1978). An analysis for transient states with application to tumor shrinkage. *Biometrics*, 34(4):571–580.

Therneau, T. and Grambsch, P. (2000). *Modeling Survival Data: Extending the Cox Model.* Springer, New York.

Tsai, W. and Crowley, J. (1998). A note on nonparametric estimators of the bivariate survival function under univariate censoring. *Biometrika*, 85(3):573–580.

Tsiatis, A. (1975). A nonidentifiability aspect of the problem of competing risks. *Proceedings of the National Academy of Sciences of the USA*, 72:20–22.

van der Vaart, A. and Wellner, J. A. (1996). *Weak Convergence and Empirical Processes. With Applications to Statistics.* Springer, New York.

van Houwelingen, H. and Putter, H. (2008). Dynamic predicting by landmarking as an alternative for multi-state modeling: an application to acute lymphoid leukemia data. *Lifetime Data Analysis*, 14:447–463.

van Walraven, C., Davis, D., Forster, A., and Wells, G. (2004). Time-dependent bias was common in survival analyses published in leading clinical journals. *Journal of Clinical Epidemiology*, 57:672–682.

Venables, W. and Ripley, B. (2002). *Modern Applied Statistics with S. 4th ed.* Springer, New York.

Wanner, C., Krane, V., März, W., Olschewski, M., Mann, J., Ruf, G., and Ritz, E. (2005). Atorvastatin in patients with type 2 diabetes mellitus undergoing hemodialysis. *New England Journal of Medicine*, 353(3):238–248.

Wolkewitz, M., Vonberg, R., Grundmann, H., Beyersmann, J., Gastmeier, P., Bärwolff, S., Geffers, C., Behnke, M., Rüden, H., and Schumacher, M. (2008). Risk factors for the development of nosocomial pneumonia and mortality on intensive care units: application of competing risks models. *Critical Care*, 12(2):R44.

Zhang, X., Zhang, M.-J., and Fine, J. (2009). A mass redistribution algorithm for right-censored and left-truncated time to event data. *Journal of Statistical Planning and Inference*, 139(9):3329–3339.

Zhang, X., Zhang, M.-J., and Fine, J. (2011). A proportional hazards regression model for the subdistribution with right-censored and left-truncated competing risks data. *Statistics in Medicine*. DOI: 10.1002/sim.4264.

Index